Contingencies, Resilience and Legal Constitutionalism

Contingency planning and resilience are of prime importance to the late modern risk society, with implications for law and for governance arrangements. Our risk society continues to seek ever more complex and detailed risk mitigation responses by law, including the UK's Civil Contingencies Act 2004 and the US Homeland Security Act 2002, which respond to counter-terrorism, natural catastrophes, and other risks.

This book seeks to analyse and criticise the legal developments in contingencies and resilience on a comparative basis, which engages not only with law and constitutionalism but also political theory and policy, including relations between public and private, national and local, and civil and military. Two transcending themes are of interest. One is institutional or structural – what bodies and power relations should we establish in a late modern world where Critical National Infrastructure is mainly held in private hands? The second is dynamic and concerns the granting of powers and arrangements for live responses. Both aspects are subjected to a strong critical stance based in 'constitutionalism', which demands state legitimacy even in extreme situations by the observance of legality, effectiveness, accountability, and individual rights.

This book was originally published as a special issue of *The International Journal of Human Rights*.

Clive Walker is Professor Emeritus of Criminal Justice Studies in the School of Law at the University of Leeds, UK. He has written extensively on terrorism and emergencies. In 2003, he was a special adviser to the UK Parliamentary select committee on the Civil Contingencies Act 2004 and subsequently published a book on that legislation.

Contingencies, Resilience and Legal Constitutionalism

Edited by
Clive Walker

Routledge
Taylor & Francis Group
LONDON AND NEW YORK

First published 2015
by Routledge

2 Park Square, Milton Park, Abingdon, Oxon OX14 4RN
711 Third Avenue, New York, NY 10017, USA

Routledge is an imprint of the Taylor & Francis Group, an informa business

First issued in paperback 2017

British Library Cataloguing in Publication Data
A catalogue record for this book is available from the British Library

ISBN 13: 978-1-138-89042-8 (hbk)
ISBN 13: 978-1-138-06001-2 (pbk)

Typeset in Times New Roman
by RefineCatch Limited, Bungay, Suffolk

Publisher's Note
The publisher accepts responsibility for any inconsistencies that may have arisen during the conversion of this book from journal articles to book chapters, namely the possible inclusion of journal terminology.

Contents

	Citation Information	vii
	Notes on Contributors	ix
1.	Legal perspectives on contingencies and resilience in an environment of constitutionalism – An overview *Clive Walker*	1
2.	Lone wolf terrorism in Norway *Catherine Appleton*	9
3.	Managing 'civil contingencies' in Australia *Michael Eburn*	25
4.	The power to react: review and discussion of Canada's emergency measures legislation *John Lindsay*	40
5.	'Keep Calm and Carry On': informing the public under the Civil Contingencies Act 2004 *Rebecca Moosavian*	59
6.	Emergency powers and the withering of the Royal Prerogative *Andrew Blick*	76
7.	The governance of emergency arrangements *Clive Walker*	92
8.	Terrorism and counterterrorism in the US: the question of responsible policy-making *John Mueller and Mark G. Stewart*	109
9.	Homeland security: definitions and accountability *Amos N. Guiora*	122
	Index	145

Citation Information

The chapters in this book were originally published in *The International Journal of Human Rights*, volume 18, issue 2 (February 2014). When citing this material, please use the original page numbering for each article, as follows:

Chapter 1
Legal perspectives on contingencies and resilience in an environment of constitutionalism – An overview
Clive Walker
The International Journal of Human Rights, volume 18, issue 2 (February 2014) pp. 119–126

Chapter 2
Lone wolf terrorism in Norway
Catherine Appleton
The International Journal of Human Rights, volume 18, issue 2 (February 2014) pp. 127–142

Chapter 3
Managing 'civil contingencies' in Australia
Michael Eburn
The International Journal of Human Rights, volume 18, issue 2 (February 2014) pp. 143–158

Chapter 4
The power to react: review and discussion of Canada's emergency measures legislation
John Lindsay
The International Journal of Human Rights, volume 18, issue 2 (February 2014) pp. 159–177

Chapter 5
'Keep Calm and Carry On': informing the public under the Civil Contingencies Act 2004
Rebecca Moosavian
The International Journal of Human Rights, volume 18, issue 2 (February 2014) pp. 178–194

Chapter 6
Emergency powers and the withering of the Royal Prerogative
Andrew Blick
The International Journal of Human Rights, volume 18, issue 2 (February 2014) pp. 195–210

Chapter 7
The governance of emergency arrangements
Clive Walker
The International Journal of Human Rights, volume 18, issue 2 (February 2014) pp. 211–227

Chapter 8
Terrorism and counterterrorism in the US: the question of responsible policy-making
John Mueller and Mark G. Stewart
The International Journal of Human Rights, volume 18, issue 2 (February 2014) pp. 228–240

Chapter 9
Homeland security: definitions and accountability
Amos N. Guiora
The International Journal of Human Rights, volume 18, issue 2 (February 2014) pp. 241–262

Please direct any queries you may have about the citations to clsuk.permissions@cengage.com

Notes on Contributors

Catherine Appleton is Senior Research Fellow at the School of Law, University of Nottingham, UK. Her research focuses on 'ultimate penalties' and the question of how societies respond to their most serious crimes. Her first monograph examined 'Life after Life Imprisonment' for a group of released life-sentenced offenders in England and Wales, and was awarded the 2011 British Society of Criminology Book Prize. She is currently engaged on a Leverhulme funded research project on Life Imprisonment Worldwide.

Andrew Blick is Lecturer in Politics and Contemporary History at the Institute of Contemporary British History, King's College London, UK. He is the author *of People Who Live in the Dark: The History of the Special Adviser in British Politics*; and a forthcoming work, *Beyond Magna Carta: A Constitution for the United Kingdom*. He has acted as a consultant on constitutional issues for organisations including the United Nations Development Programme and the UK National Audit Office; and undertaken research projects for the Rowntree Trust, Carnegie UK, the German Marshall Fund, and the Madison Trust.

Michael Eburn is Associate Professor in the College of Law at the Australian National University, Canberra, Australia. He is the author of numerous journal and conference papers in law and emergency management as well as the book, *Emergency Law* (2009; 4th ed. forthcoming). His PhD (2009) benchmarked Australian law against the guidelines published by the International Red Cross to facilitate international disaster assistance. He is currently the Chief Investigator and Project Leader on a research project, to be funded by the Bushfire and Natural Hazards CRC, exploring 'Policies, Institutions and Governance of Natural Hazards'.

Amos N. Guiora is Professor of Law and Co-Director of the Center for Global Justice, S. J. Quinney College of Law, University of Utah, Salt Lake City, USA. His recent books include *Homeland Security: What Is It and Where Are We Going?* (2011); *Freedom from Religion: Rights and National Security* (2009, 2012); *Legitimate Target: A Criteria based Approach to Targeted Killing* (2013); *Modern Geopolitics and Security: Strategies for Unwinnable Conflicts* (2013); and *Tolerating Intolerance: The Price of Protecting Extremism* (2013).

John Lindsay is Associate Professor in the Department of Applied Disaster and Emergency Studies at Brandon University, Manitoba, Canada, and a candidate for the PhD in Emergency Management at the Joint Centre for Disaster Research, Massey University, Palmerston North, New Zealand. He integrates 20 years of experience as an emergency management professional into his research focus on urban planning, hazard mitigation and emergency management policy.

Rebecca Moosavian is Senior Lecturer in the School of Law at Northumbria University, Newcastle, UK, and teaches Jurisprudence and Intellectual Property Law across a range of undergraduate and postgraduate courses. She acted as course leader for the Open Learning LL.B programme from 2004 to 2010. In 2009 she completed an M(Phil) thesis on prime ministerial exercise of the war prerogative in the Iraq war. She has published work in the areas of constitutional law and privacy, and her wider research interests include intellectual property and jurisprudence, particularly critical legal theory.

John Mueller is the Ralph D. Mershon Senior Research Scientist at the Mershon Center for International Security Studies, and Adjunct Professor of Political Science, Ohio State University, Columbus, OH, USA. He is also a Senior Fellow at the Cato Institute, Washington, DC, USA. He is a member of the American Academy of Arts and Sciences, has been a John Simon Guggenheim Fellow, and has received grants from the National Science Foundation and the National Endowment for the Humanities. His book, *Terrorism, Security, and Money: Balancing the Risks, Benefits, and Costs of Homeland Security* (with Mark Stewart) was published in 2011. Other recent books include *Atomic Obsession: Nuclear Alarmism from Hiroshima to Al Qaeda* (2010) and *War and Ideas: Selected Essays* (2011).

Mark G. Stewart is Professor of Civil Engineering and Director of the Centre for Infrastructure Performance and Reliability, University of Newcastle, Australia. He is the co-author of *Terror, Security, and Money: Balancing the Risks, Benefits, and Costs of Homeland Security* (2011), and has produced over 350 technical papers and reports. He has more than 25 years of experience in probabilistic risk and vulnerability assessment of infrastructure and security systems that are subject to man-made and natural hazards. In 2011 he received a five-year Australian Professorial Fellowship from the Australian Research Council to develop probabilistic risk-modelling techniques for infrastructure subject to military and terrorist explosive blasts, and cost–benefit assessments of counter-terrorism protective measures for critical infrastructure.

Clive Walker is Professor Emeritus of Criminal Justice Studies at the School of Law, University of Leeds, UK. He has written extensively on criminal justice and human rights issues. His books have focused upon terrorism and the law, including, *Terrorism and the Law* (2011). In 2003, he was a special adviser to the UK Parliamentary select committee considering the draft proposals which emerged as the Civil Contingencies Act 2004. A book commentating upon that Act, *The Civil Contingencies Act 2004: Risk, Resilience and the Law in the United Kingdom*, was published in 2006. Other recent publications include: *Free Speech in an Internet Era* (with Russ Weaver, 2013); *Counter-Terrorism, Human Rights and the Rule Of Law: Crossing Legal Boundaries in Defence of the State* (co-edited with Aniceto Masferrer, 2013); *Dirty Assets: Emerging Issues in the Regulation of Criminal and Terrorist Assets* (with Colin King, 2014) and *The Anti-Terrorism Legislation* (Third edition, 2014).

Legal perspectives on contingencies and resilience in an environment of constitutionalism – An overview

Clive Walker

School of Law, University of Leeds, UK

Contexts

Many of the hazards faced by twenty-first century society reflect changes within the broader risk domain which are characterised by high levels of unpredictability,[1] and low tolerance of risk.[2] By these trends there emerges a 'risk society'.[3] In turn, the crises caused by the actual occurrence or simply the threat of terrorism, pandemic human and animal diseases, and climate disasters such as hurricanes and floods have all been taken to be reflexive outcomes of late modernity, which were much as predicted by the risk theorists. Thus, late modern terrorism can be interpreted as an unwanted by-product of networked modes of relationship, the removal of sovereign borders, and even the end of dialectical political progress in which we instead witness 'triumphant globalisation fighting with itself'.[4] Likewise pandemic diseases and disasters are posited as the darker sides of scientific or industrial 'progress'.[5]

Such hazards seem to demand novel responses and regulations which can take account of the post-industrial landscape and the 'hollowed out' state.[6] While the role of government remains that of titular guarantor of security for its citizens,[7] a task which the state still relishes since it accords legitimacy and allegiance, the mission cannot be accomplished by government alone, but in a late modern sense is delivered by networked modes including private stakeholders and indeed the mobilisation of all citizens to practice self-resilience. Thus, for the effective management of risk, and for resilience and recovery to be achieved, participation by a range of public and private sector stakeholders is required but can be obdurately difficult to deliver.[8] In an ideal scenario these networks improve the stability of policy-making around contingencies and increase the likelihood of a policy objective being achieved on account of the wider processes of information-gathering and risk-assessment. However, in seeking to provide stability, government intervention may embody the potential for unintended consequences, including financial and political costs, as well as negative impacts in terms of displacement.[9] These divergent results reflect the dynamic and complex nature of risk and security management in late modernity and the key challenges facing institutional actors, such as uncertainty and the limits of knowledge. Policy communities may also impose constraints on the way pertinent information is communicated through the pre-filtering of information or the implementation. These problems, in turn, may adversely affect the way public sector institutions respond to, or prepare for, crisis, particularly in the context of a networked governance framework which is oriented around resilience.

The collection of papers in this special issue will not provide a blueprint for overcoming these manifold woes of the late modern conceptualisation and management of risk, but they will provide original and insightful commentaries which explain and analyse the problems arising, and offer some conclusions as to how some of the difficulties of emergency response can be overcome. There are two important contexts within which the contributors have been asked to undertake this work.

One is that we have sought to adopt a 'legal perspective', by which is meant that the papers in the collection focus upon legal mechanisms to build, structure or exercise power. It is notable that law has played a prominent role as an instrument in contingencies management and emergency response in recent times, for it seems contrary to some of the discourses of late modernity, wherein national sovereignty is downplayed. Exploring this tension, we have chosen to concentrate on legal mechanisms at the national level, since much of the effective action is still taken at that level when issues of national security are at stake – a domain which is one of the vestiges of old-time sovereignty, where states are often much less willing to share than with other policies. The partial coverage and uneven implementation of European Union involvement in security affairs is testament to sovereign reticence,[10] although progress has been made with contingency planning in the field of major critical infrastructure,[11] and on a broader front with national risk assessments[12] and two Council Decisions on the Civil Protection Mechanism, which facilitate cooperation between the Member States and the Union in the field of civil protection.[13] In any event, the focus on the legal is not intended to mean just the core instruments of legislation and judicial decisions. Rather, plenty of materials will be found in the papers in this collection where states and other authorities have resorted to a much wider range of instruments, especially guidance, mutual understandings and other forms of governance devices.

The second parameter is 'Constitutionalism', which provides a critical perspective for the papers in the collection. As with the 'legal perspective', the concept is not defined narrowly but is understood as the collection of values which foster state legitimacy, even in extreme situations, and principally comprising: transparency and legality; effectiveness and accountability; and respect for individual rights.[14] This focus is again apparently paradoxical. The most notable emergency of recent times, the terrorism emergency of 9/11 and thereafter, has, it is claimed, been met not only by bloated security laws and apparatus but also by the rising assertion of judicial and parliamentary review. This trend towards greater assertiveness in security disputes challenges the assertion that anti-terrorism legislation represents naked power – rule by law and not by the rule of law.[15] A more principled and controlled approach towards anti-terrorism legislation is far from that of a totalitarian state of exception which operates in a legal vacuum.[16] Thus, most governments (with one notable exception) have not declared an all-encompassing 'war on terror' which rivals or replaces regular laws, nor have they engendered a 'zone of anomie in which all legal determinations are deactivated'.[17] Those commentators who peddle such a picture appear ignorant of the efforts of the international community since 1945 to determine that no such state of anomie or Hobbesian state of savagery can be authorised again under the contemporary codes of international humanitarian laws and international human rights laws. The contemporary rulers of liberal democracies who have sought to draw upon the poisonous old doctrines of absolute executive authority have rightly faced concerted legal opposition. Most norms have not at any time been suspended, as with the 'war on terror' or 'criminal law of the enemy',[18] but societies have adapted to a new normality of terrorism. As the UK Foreign Secretary, David Miliband, declared, 'We must respond to terrorism by championing the rule of law, not by subordinating it, for it is the cornerstone of the democratic society.'[19] Even US President Obama has consistently called for an end to exceptional measures (such as detention at Guantánamo Bay),[20] although his championing of drone strikes may be no less abusive of the rule of law.[21] The trend towards constitutionalism demonstrates that its principles are becoming 'not a mask but the true image of our nation',[22] and offer protection from naked state power.[23] The law has thereby demonstrated capacity through its normative power and through the independence of its institutions to respond to executive claims to power even in circumstances of dire emergency, and to impose some ethical restraint.[24] The papers in this collection will consider the success of these ethical constraints.

The papers

There are eight papers in this special issue. They might be categorised in various ways, but the following four part typology seems the strongest to emerge.

The first theme concerns the meanings of contingency and emergency. The wider the categorisation, the more that 'special' measures are needed (and perhaps the more special they become). This net-widening process is demonstrated in various jurisdictions. In the UK the Civil Contingencies Act 2004 imports a far wider definition of 'emergency' in sections 1 and 19 than the legislation it replaced, which was essentially concerned with mass industrial dislocation.[25] This legislation is covered in the papers of Moosavian and Walker, but it is the paper of Catherine Appleton, 'The Breivik Report: Lessons from Norway's "Lone Wolf" Terrorist Attacks', which sheds novel light on the struggle to place boundaries around the conception of emergency and the state's claim to take exceptional action. Her paper considers Anders Behring Breivik's massacre of 77 people at two sites in Norway on 22 July 2011 and assesses fully the dangers and potential dangers posed by 'lone wolf' or lone operator terrorist acts. As demonstrated on 22 July, the lone wolf operational model presents a number of critical challenges for crisis management and emergency preparedness, even for highly ordered and well-financed countries such as Norway. According to Bakker and de Graaf, 'lone wolf terrorist acts should be considered "black swan" occurrences that are almost impossible to categorize or systematize, let alone forecast'.[26] The aim of Appleton's paper is to draw attention to the phenomenon of lone wolf terrorism and to critically assess the response by the Norwegian authorities to the twin terror attacks in 2011. In so doing, it discusses the implications of lone wolf terrorism for crisis management and contingency planning, and argues that, while law enforcement plays a central role in efforts to combat lone wolf attacks, it is essential that counterterrorist approaches be based on democratic principles and respect for human rights.

The second theme concerns the allocation of emergency powers and responses, which can be conceived as involving localised–centralised, open or closed, or micro, meso and macro power relationships. This problem is certainly in play in the UK and US. In the UK, the Civil Contingencies Act 2004 provides some clarity by specifying resilience planning and response as being primarily a local or devolved (Scottish and Welsh) responsibility under Part I, whereas action in the event of serious emergency under Part II is for national leadership. At the same time, there are overlaps and abeyances which are not explained. In the US, there is the added complexity of a federal system of government, which can have the effect of leaving responsibility unallocated, as occurred in the case of Hurricane Katrina in 2004, when the mantle of 'all-risks' administrator conferred on the Department of Homeland Security (DHS) proved tragically threadbare.[27] Unfortunately it is during a situation of emergency that the interfaces between the nodes of state and the limits of its competences are laid bare. As well as in the papers touching on these jurisdictions, the issue of allocation is especially highlighted for Australia and Canada, respectively.

Michael Eburn deals with 'Legal Issues and Civil Contingencies in Australia: All Hazards? All Agencies?'. As well as describing overall how 'civil contingencies' are tackled in the country, the paper plots the allocation of responsibilities for emergency management across Australia's federated system of government and provides a critical review of the institutional and structural arrangements. It is argued that, even though Australia's emergency management arrangements have generally worked well, there are significant gaps because of the unevenness of allocation. Limitations include the absence of any established Commonwealth response powers and weak arrangements to coordinate the Commonwealth response to emergency. It is further argued that Australia's arrangements focus on official responses to emergencies (at whatever level). They have, to date, largely omitted other players, in particular non-government actors, whether private industry, communities or individuals.

John Lindsay considers 'Disaster Management in Canada'. The story begins with the War Measures Act of 1914, which conferred on the federal government extraordinary powers to deal with the domestic implications of the First World War. The government further expanded and institutionalised these powers during the Second World War, the Korean crisis and the Cold War. These war emergency powers are now entrenched in both federal and provincial emergency measures legislation and can be exercised for peacetime disasters. Lindsay's paper reviews the historical progression of these powers to better understand the original context for their development and to connect past abuses of civil rights under the War Measures Act to the protections and limits incorporated in the current legislation. Then, based on this understanding of where the powers originated, his paper takes a comprehensive look at the powers available across the Canadian jurisdictions. It identifies common groups of powers and discusses regional variations. It concludes with a short discussion on the imbalance between extraordinary powers for response and the lack of a corresponding emphasis on reducing the risks that may generate the need for them.

Another way of considering the allocation of powers is taken up by Rebecca Moosavian in 'Power, Knowledge and Part 1 of the Civil Contingencies Act 2004'. Here she is asking whether the legislation's ideal of energising stakeholders to plan for their own resilience and responses in the face of potential emergencies is actually delivered. Part 1 of the Civil Contingencies Act 2004 provides a framework that governs the planning and preparations for a wide range of potential emergencies. The Act explicitly addresses a range of 'Responders' within the public government or utility sectors, but it says little about empowering or at least involving the individual citizen. It is apparent that information plays a fundamental role in the Part 1 planning provisions and associated guidance. Her article focuses on one specific information-related provision, namely, the duty to maintain plans to warn and inform the public in the event of an emergency. It undertakes detailed analysis of the legislative provisions and extensive government guidance regarding informing the public and related media-handling. It analyses these measures in light of two competing organisational models – traditional authoritarian 'command' structures, on the one hand, and more flexible, decentralised arrangements on the other.

A third issue should be theme in this collection relates to the instruments of power. What forms of power are most suitable for responding to emergencies? This question touches upon the second issue of the level of response – local or central and also ministerial or parliamentary. It also engages the next issue, ethical compliance, since one format of instrument may better embody attributes of constitutionalism, such as certainty and accountability, than another. The topic is taken up here in two papers.

One is delivered by Andrew Blick, who addresses the evolution of emergency powers in 'Emergency Powers and the Withering of the Royal Prerogative'. He notes that a feature common to most governments in different time periods and places is that they possess exceptional powers applicable to emergencies. These authorities are necessary but vulnerable to abuse. Democracies seek to create and restrain such powers through constitutional texts and supplementary legislation. In the UK, lacking a written constitution, many of these executive entitlements have traditionally derived from the Royal Prerogative. This constitutional source comprises a group of residual monarchical authorities that have largely devolved in practice to ministers. As an historical entity, the prerogative can only fully be understood in its more recent manifestations through taking a long-term perspective. It has often attracted controversy on the grounds that it is not subject to sufficient levels of what would now be termed accountability. At various points, reform has been demanded and sometimes achieved. Since the later decades of the 20th century, there has been an erosion of the prerogative, including its emergency powers facets. Parts of it have transferred to a statutory basis, while those elements that remain have become subject to enhanced legal and political constraints. This trend was part of a wider process of constitutional transformation traceable to the 1950s. Early in the twenty-first century the Royal Prerogative, already undergoing change, moved further to the centre of the constitutional reform agenda. This heightened salience arose from altered global security concerns, and associated developments in UK external policy. The Iraq conflict especially encouraged a focus on the powers under which the UK participated in this action. Calls for change, aimed at giving Parliament an enhanced role in war-making and over other matters, followed. Eventually a broad government reform package appeared. A convention regarding parliamentary consultation over engagement in armed combat developed. But features of the prerogative have persisted, including those allowing for government to act in the most perilous of circumstances. A thorough eradication of the Royal Prerogative could require a more complete overhaul of UK governmental arrangements, embodying them in a codified written constitution.

Next, Clive Walker considers 'The Governance of Emergency Powers' very much within the context of Part II of the Civil Contingencies Act 2004. These particular 'emergency powers' have remained almost a dead-letter since enactment and have never been fully implemented. The first part of the paper will reflect upon the contents and performance of Part II. The second will seek to explain the powerful official aversion to Part II as an instrument of empowerment. The reasons for this aversion relate not only to the political implications arising from the implied admission of a loss of control or failure of resilience but also to a disinclination to comply with the mechanisms for limitation and accountability in Part II. The latter point is evidenced by the production of draft Bills which set a lower threshold for use whenever a suitable type of emergency arises. Two such Bills have appeared: the Draft Detention of Terrorist Suspects (Temporary Extension) Bill 2011 and the draft Enhanced Terrorism Prevention and Investigations Measures Bill 2011. This growing trend will be analysed and criticised with respect to the potentially diminished governance of resilience and emergency measures.

The final theme in this special collection reflects our emphasis on the pursuit of 'constitutionalism'. This issue as it pertains to individual rights has already been addressed in some of the foregoing papers, especially those by Lindsay and Walker. Two further papers focus upon accountability and effectiveness in political and economic aspects.

John Mueller and Mark Stewart take up this debate in 'Terrorism and Counterterrorism in the US: Threat, Capacity, Risk, Cost, Benefit'. Officials serving the public are tasked at the most fundamental level to spend funds in a manner that most effectively and efficiently keeps people safe. To do otherwise is irresponsible and, because human lives are at stake, immoral. In the case of counterterrorism policy-making, it is important, then, to evaluate the degree to which any gains in security afforded by counterterrorism measures have been great enough to justify their cost. Risk analysis is an aid to responsible decision-making that does exactly that. Four issues are considered central to this approach, and they are applied to the hazard presented by terrorism: the cost per saved life, acceptable risk, cost–benefit analysis, and risk communication. The paper also assesses the degree to which risk analysis has been coherently applied to counterterrorism efforts in the US in making or evaluating decisions that have cost taxpayers many hundreds of billions of dollars over the past dozen years.

Finally, Amos Guiora, in 'Homeland Security: Accountability, Effectiveness and the Price to Pay' addresses two core issues relevant to the broader homeland security discussion: framing homeland security; and articulating criteria for effectiveness and accountability in Homeland Security. In the light of the Bush administration's forays into homeland security, the paper seeks to articulate the basis for the creation of the DHS in the aftermath of 11 September 2001. Therefore, the focus will be on the past; while, perhaps, shedding light on the present and the future, the primary purpose is a reflection on Bush administration policies. In large part the Obama administration has not made overwhelming changes to the philosophical underpinnings of the DHS. While the challenges have changed, the essence of DHS is largely similar between the two administrations. It is for this reason, then, that in addressing Homeland Security, this paper takes a look back.[28]

The wider project

This collection arises from a wider seminar series, 'In Search of Resilience: Exploring Shifting Paradigms of Contingency Management', which was funded by the ESRC and under the direction of Martina McGuinness.[29] The aim of this seminar series was to provide a forum for policy-makers, practitioners and academics to critically review and debate shifts in contingency management and the implications for building resilience. The current collection is based around the papers submitted to the symposium on 'Legal Perspectives on Contingencies and Resilience', held at the Royal United Services Institute, London, on 19 March 2013.[30]

The processes of drafting and applying laws such as the Civil Contingencies Act 2004 to contingencies and resilience in an environment of constitutionalism have shed light on hitherto closed practices and relationships. Given the amount of effort now expended on planning and risk aversion, it might be seen as almost a failure if an actual emergency actually occurs, although, as argued by Aaron Wildavsky,[31] there is much to learn also from error and how to recover in stronger shape. The papers in this special issue seek to draw lessons from all phases of response to emergency and from the ethical considerations which should overlay any response.

Notes

1. See Michael Power, *Organized Uncertainty: Designing a World of Risk Management* (Oxford: Oxford University Press, 2007).
2. Frank Furedi, *Culture of Fear: Risk Taking and the Morality of Low Expectation*, revised ed. (London: Continuum, 2002).
3. Uncertainty extends to the existential issue of the categories to be covered. Enrico L. Quarantelle, ed., *What is a Disaster?* (New York: Routledge, 1998).
4. Jean Baudrillard, 'The Spirit of Terrorism' (2001): 11, Leuk-Stadt: http://www.egs.edu/faculty/jean-baudrillard/articles/the-spirit-of-terrorism/. See also Francis Fukuyama, *The End of History and the Last Man* (London: Penguin, 1992).

5. Ulrich Beck, *Risk Society* (London: Sage, 1992); Beck, 'Risk Society and the Provident State', in *Risk, Environment and Modernity*, ed. Scott Lash, Bronislaw Szerszynski and Brian Wynne (London: Sage, 1998); and Beck, *World at Risk* (Cambridge: Polity, 2009).
6. See Rod A.W. Rhodes, *Understanding Governance: Policy Networks, Governance, Reflexivity and Accountability* (Buckingham: Open University Press, 1997), 17.
7. See, for example, *McCann v. United Kingdom*, App no 18984/91, Ser A 324 (1995); *López Ostra v. Spain*, App no 16798/90, 12 September 2009; and *Kolyadenko v. Russia*, App nos 17423/05, 20534/05, 20678/05, 23263/05, 24283/05 and 35673/05, 28 February 2012.
8. Michael Pitt, *Learning Lessons from the 2007 Floods*, First Report (London: Cabinet Office, 2007), 77; House of Commons Select Committee on Environment, Food and Rural Affairs, *Flooding* (2007–8 HC 49–1); House of Commons Environment, Food and Rural Affairs Select Committee, *Managing Flood Risk* (2013–14 HC 330); and Jill Morgan and Mark Stalworthy, 'Indemnifying against Flood Loss in a Changing Environment', *Legal Studies* 33 (2013): 239.
9. The creation of a 'Ring of Steel' in one place may result in the choice of easier targets rather than desistance. Clive Walker, 'Political Violence and Commercial Risk', *Current Legal Problems* 56 (2004): 531; Stephen Graham, ed., *Cities, War and Terrorism* (Oxford: Blackwell, 2004); and Tony Craig, 'Sabotage!', *Intelligence and National Security* 25 (2010): 309.
10. See Jörg Monar, 'Common Threat and Common Response? The European Union's Counter-terrorism Strategy and its Problems', *Government and Opposition* 42 (2007): 292; Monica Den Boer, Claudia Hillebrand and Andreas Nölke, 'Legitimacy under Pressure: The European Web of Counter-terrorism Networks', *Journal of Common Market Studies* 46 (2008): 101; Oldrich Bures, *EU Counterterrorism Policy: A Paper Tiger?* (Farnham: Ashgate, 2011); and Cian Murphy, *EU Counter-terrorism Law* (Oxford: Hart, 2012).
11. Council Directive 2008/114/EC of 8 December 2008 on the identification and designation of European critical infrastructures and the assessment of the need to improve their protection. The Directive is set in the context of a Council Communication on a European Programme for Critical Infrastructure Protection (COM(2006) 786 final) and follows a European Commission Green Paper with the same title (COM(2005) 576 final). See Bernhard Hämmerli and Andrea Renda, *Protecting Critical Infrastructure in the EU* (Brussels: Centre for European Policy Studies, 2010); Javier Lopez, Roberto Setola and Stephen Wolthusen, eds., *Critical Infrastructure Protection: Advances in Critical Infrastructure Protection* (Heidelberg: Springer, 2012); and Theodore Konstadinides, 'Civil Protection in EU Law', *European Law Journal* 19 (2013): 267. For UK criticism, see House of Commons European Scrutiny Committee, *Fifteenth Report* (2006–07 HC 41-xv).
12. See European Commission, *Risk Assessment and Mapping Guidelines for Disaster Management*, SEC(2010) 1626 final (Brussels: European Commission, 2010).
13. Council Decision 2007/779/EC, Euratom of 8 November 2007 establishing a Community Civil Protection Mechanism (recast); and Council Decision 2007/162/EC, Euratom of 5 March 2007 establishing a Civil Protection Financial Instrument. See, further, Proposal for a Decision of the European Parliament and of the Council, *Union Civil Protection Mechanism*, COM(2011) 934 final (Brussels: European Union, 2011); and Council of the European Union, *Joint Proposal for a Council Decision on the arrangements for the implementation by the Union of the Solidarity Clause*, JOIN(2012) 39 final (Brussels: Council of the EU, 2012).
14. For its application to terrorism emergencies, see Clive Walker, 'Constitutional Governance and Special Powers against Terrorism', *Columbia Journal of Transnational Law* 35 (1997): 1; and Walker, 'Cosmopolitan Liberty in the Age of Terrorism', in *International and Comparative Criminal Justice and Urban Governance*, ed. Adam Crawford (Cambridge University Press, Cambridge, 2011), 413.
15. See Giorgio Agamben, *The State of Exception* (Chicago: University of Chicago Press, 2005).
16. Compare Carl Schmitt, *Political Theology* (Cambridge: MIT Press, 1985).
17. Agamben, *The State of Exception*, 50.
18. See Günther Jakobs and Cancio Meliá, *Criminal Law of the Enemy*, 2nd ed. (Madrid: Civitas, 2006).
19. In the *Guardian*, January 15, 2009, 29.
20. See Executive Order 13492 – Review and Disposition of Individuals Detained at the Guantánamo Bay Naval Base and Closure of Detention Facilities (22 January 2009); and Executive Order 13493 – Review of Detention Policy Options (22 January 2009). The

policy of closure is effectively blocked by Congress, beginning with the National Defense Authorization Act for Fiscal Year 2011, s.1032 (PL 11–383).

21. See Christof Heyns, Special Rapporteur on Extrajudicial, Summary or Arbitrary Executions, A/68/382 (New York: United Nations, 2013); and Special Rapporteur on the Promotion and Protection of Human Rights and Fundamental Freedoms while Countering Terrorism, A/68/389 (New York: United Nations, 2013).
22. Michael Ignatieff, *The Lesser Evil* (Edinburgh: Edinburgh University Press, 2005), 144.
23. See Sarah Marks, 'State Centrism, International Law and the Anxieties of Influence', *Leiden Journal of International Law* 19 (2006): 339; and Barry Vaughan and Sean Kilcommins, *Terrorism, Rights and the Rule of Law* (Cullompton, UK: Willan, 2008), 13.
24. See Colm Campbell and Ita Connelly, 'Making War on Terror?', *Modern Law Review* 69 (2006): 935, 939.
25. The agenda of 'protect' in counter-terrorism strategy is also hugely expansive. See Clive Walker, '"Protect" against Terrorism: In Service of the State, the Corporation, or the Citizen?', in *The Long Decade: How 9/11 has Changed the Law*, ed. David Jenkins, Anders Henriksen and Amanda Jacobsen (Oxford: Oxford University Press, 2014).
26. Edwin Bakker and Beatrice de Graaf, 'Preventing Lone Wolf Terrorism: Some CT Approaches Addressed', *Perspectives on Terrorism* 5, nos. 5–6 (2003): 43.
27. See Select Bipartisan Committee to Investigate the Preparation for and Response to Hurricane Katrina, *A Failure of Initiative* (Washington, DC: 2006); and Daniel A. Farber and Jim Chen, *Disasters and the Law: Katrina and Beyond* (Frederick, MD: Aspen, 2006).
28. See also Thomas Cmar, 'Office of Homeland Security', *Harvard Journal on Legislation* 39 (2002): 455; and Jonathan Thessin, 'Department of Homeland Security', *Harvard Journal on Legislation* 40 (2003): 513.
29. RES-451–26-0901.
30. Thanks for providing the venue (and a paper on 'Local Resilience and National Risk') to Jennifer Cole, Senior Research Fellow, Resilience and Emergency Management at the Royal United Services Institute (RUSI).
31. Wildavsky, *Searching for Safety* (Piscataway, NJ: Transaction Publishers, 1988).

Lone wolf terrorism in Norway

Catherine Appleton

School of Law, University of Leeds, UK

Anders Behring Breivik's massacre of 77 people on 22 July 2011 sheds light once again on the dangers and potential dangers posed by 'lone wolf' or lone operator terrorist acts. As demonstrated on 22 July, the lone wolf operational model presents a number of critical challenges for crisis management and emergency preparedness. The aim of this article is to draw attention to the phenomenon of lone wolf terrorism and to critically assess the response by Norwegian authorities to the twin terror attacks in 2011. In so doing, it will discuss the implications of lone wolf terrorism for crisis management and contingency planning, and argue that, while law enforcement plays a central role in efforts to combat lone wolf attacks, it is essential that counterterrorist approaches be based on democratic principles and respect for human rights.

Introduction

On 22 July 2011 Norway was struck by two terror attacks of unprecedented magnitude for that country. First, an improvised explosive device hidden in a Volkswagen Crafter van was detonated at the centre of the executive government quarter in the capital of Oslo, destroying several central buildings, including the office of the prime minister. Eight people were killed and another nine were seriously injured. Less than two hours after the explosion a gunman disguised as a police officer opened fire on a summer camp organised by the Norwegian Labour Party's youth association, AUF, on the island of Utøya, located 32 kilometres outside Oslo. The shooting claimed the lives of 69 individuals; many others were injured in the attack. Most of the victims were aged between 15 and 18 years old. The perpetrator, 32-year-old Anders Behring Breivik, was arrested the same evening and immediately confessed to committing both attacks. Breivik, a Norwegian citizen, was charged under criminal law for acts of terrorism and premeditated murder. A year later the Oslo District Court established that he had operated alone. He was declared sane, found guilty and sentenced to Norway's maximum sentence: at least 21 years' preventive detention (*forvaring*) which, under Norwegian law, can be repeatedly extended for five years as long as he is still considered a danger to society.[1]

The recent terrorist attacks in Norway have heightened the perceived threat of lone wolf terrorism – or terrorist acts carried out by single individuals – among Western nations. Less than a month after Breivik was charged with the killings, President Barak Obama announced that 'the biggest concern we have right now . . . is the lone wolf terrorist, somebody with a single weapon being able to carry out wide-scale massacres of the sort that we saw in Norway recently'.[2] Australian Security Intelligence Organisation Director-General David Irvine joined the list of concerned officials when he stated that an event similar to the terrorist attacks in Norway could not be ruled out. Irvine commented that 'particular worries are the so-called "lone wolf" or "stand-alone" groups who act independently and throw off few clues as to malicious intent'.[3] Furthermore, European security officials have recently argued that the issue of lone wolf terrorism requires increasing attention. Europol's 2012 situation and trend report on terrorism noted that 'serious threats emanate not only from established terrorist organisations but increasingly from lone actors . . . whose radicalisation takes place largely undetected'.[4]

Despite the increased concern surrounding the threat of solo-actor terrorist attacks, limited scholarly attention has focused on the concept of lone wolf terrorism, or explored the broader implications for crisis management or contingency planning.[5] This article aims to explore the phenomenon of lone wolf terrorism and discuss the challenges that the 22 July attacks presented for crises and contingencies management in Norway. First, it aims to problematise the definition of lone wolf terrorism and briefly explore the evolution and increase of such terrorist attacks. Second, it will investigate the events that took place in Oslo and Utøya, drawing on the findings of the 22 July Commission report ordered by Norway's parliament following the attacks. In so doing, the article will discuss the implications of lone wolf terrorism for crisis management and contingency planning, and critically assess counterterrorist approaches to the threat posed by lone wolf terrorism.

Defining lone wolves

There is a small but growing body of research focusing on acts of terrorism by single individuals or so-called lone wolves.[6] Yet no academic consensus exists regarding the definition of lone wolf terrorism. In one of the major works on this topic, Simon defines lone wolf terrorism as 'the use or threat of violence . . . by an individual acting alone or with minimal support from one or two other people'.[7] Similarly Hewitt's analysis of incidents of lone wolf terrorism in the US between 1955 and 1999 includes acts committed by terrorist groups of up to four people.[8] But clear objections have been raised on the grounds that lone terrorists and group terrorism are not the same.[9] Conversely Burton and Stewart argue that a lone wolf 'is a person who acts on his or her own without orders from – or even connections to – an organization'.[10] They distinguish lone wolves from sleeper cells, arguing that sleepers are operatives who infiltrate the targeted society or organisation until being activated by a group or organisation. For Burton and Stewart a lone wolf is a 'stand alone operative who by his very nature is embedded in the targeted society and is capable of self-activation at any time'.[11] By emphasising the absence of connections with a broader network or organisation, however, Burton and Stewart fail to consider the ideological connections individuals might have with other networks or organisations, either through personal connections, inspirational material or via the internet.[12]

This issue has been addressed clearly in Spaaij's recent conceptualisation of lone wolf terrorism.[13] According to Spaaij, lone wolf terrorism involves terrorist attacks carried out by persons who: '(a) operate individually, (b) do not belong to an organized terrorist group or network, and (c) whose modi operandi are conceived and directed by the individual without any direct outside command or hierarchy'.[14] This definition includes such

Islamist lone wolves as Abdulhakim Mujahid Muhammad, who opened fire on a US military recruiting office in Little Rock, Arkansas, right-wing extremists like David Copeland, who became known as the 'London Nail Bomber' after a two-week bombing campaign in April 1999 targeting London's Black, Asian and gay communities, as well as Eric Rudolph, the 'Olympic Park Bomber', who perpetrated a series of bombings across the southern US between 1996 and 1998 that claimed three lives and wounded more than 120 others. But, as noted by Spaaij, it excludes certain high-profile terrorist attacks that are commonly ascribed to lone individuals, most notably the Oklahoma City bombing on 19 April 1995. Although the attack was carried out by an individual, Timothy McVeigh, his accomplice Terry Nichols appears to have played a considerable role in preparations for the attack. A group is thus considered to comprise two or more people.

Importantly Spaaij's interpretation of lone wolf terrorism allows for a degree of external influence during the so-called terrorist cycle. Lone wolves are often linked with, and influenced by, individuals or organisations whose ideologies and beliefs they support. Indeed, Pantucci, in his 'typology of lone wolves', argues that 'lone wolves . . . are individuals who, while appearing to carry out their actions alone and without any physical outside instigation, in fact, demonstrate some level of contact with operational extremists.'[15] This raises a crucial terminological issue for Gerry Gable and Paul Jackson in their recent report, *Lone Wolves: Myth or Reality?*. They conclude that 'far-right terrorists are not lone wolves but are connected with, influenced by and often helped by organisations whose beliefs they share'.[16] Gable and Jackson point out that the idea of lone wolf terrorism is now applied to a wide range of cases of solo-actor terrorists. The authors comment that, while 'there is good reason to believe that the most vital components of their activity were solo ventures . . . closer scrutiny shows not only that the solo actor in each case was motivated by a wider cause, but that each terrorist actor believed their actions would benefit a wider community, whom they were acting "for"'.[17] For Gable and Jackson the casual use of the term 'lone wolf' fails to grasp 'both the particular context from which "lone wolf" ideology comes, and the community of support that backs up such solo actor terrorism'.[18] Feldman concurs with this assessment but also acknowledges that, while self-activating terrorists like Breivik do not operate in a social vacuum,[19] this should not disqualify use of the expression 'lone wolf terrorism'; rather, it should 'properly add nuance to constructions of this phenomenon'.[20]

Attempting to define acts of terrorism has always been difficult, in part because the concept is dynamic.[21] The lack of consensus among academics and commentators on lone wolf terrorism highlights the intrinsic definitional difficulties of the term, and raises serious questions regarding the validity and generalisation of academic studies. With these limitations in mind, the next section briefly considers the emergence and prevalence of the lone wolf phenomenon.

The rise of lone wolf terrorism

Solo acts of terrorism are not a recent phenomenon. Acts of terrorism committed by single individuals who operate outside established terror networks have been traced back to 19th century anarchists who staged political assassinations and bombings under the strategy of 'propaganda by deed', as promulgated by 19th century Russian anarchist Mikhail Bakunin (1814–76). Bakunin's writings urged individuals to kill those who represented an existing social order, and inspired terrorist attacks in many parts of the Western world.[22] After the Second World War the concept of solo actor terrorism gained momentum among white supremacists and anti-government extremists in the US, and was associated particularly with the strategy of 'leaderless resistance'.[23] Kaplan defines leaderless resistance as a 'lone wolf operation in which an individual, or a very small, highly cohesive group, engages in acts of anti-state violence independent of any movement, leader, or network of support'.[24] The concept of leaderless resistance was popularised during the 1980s and 1990s by white supremacist Louis Beam, a former Ku Klux Klan and Aryan Nations member. Beam's vision was one whereby 'all individuals and groups operate independently of each other, and never report to a central headquarters or single leader for direction or instruction'.[25]

In the same vein white supremacists Tom Metzger and Alex Curtis promoted the term 'lone wolf' in the late 1990s in order to popularise the image of the lonesome, 'patriotic' warrior.[26] They envisioned lone wolves to be considerably more difficult to detect than conventional group-actor terrorists. Indeed, Alex Curtis encouraged fellow racists to act alone in committing violent crimes so that they would not implicate others. Both men recognised the opportunities the internet offered for the circulation of information and for communication with fellow assailants. The recent terror attacks in Norway resonate with their ideas and exemplify how the internet has become a convenient means for disseminating radical material. In the hours before his attacks on Oslo and Utøya Breivik emailed a copy of his 1518-page manifesto to several thousand recipients and posted the document – *2083: A Declaration of European Independence* – on the internet. Breivik's manifesto provided a detailed description and diary of his operations, justifying the massacre as a necessary measure to save Norway and Europe from Muslim immigrants, and specifically to punish his country's government for embracing European multiculturalism.[27]

Although not a new phenomenon, the emergence of the leaderless lone wolf terrorist has recently been portrayed in the media as a growing security threat.[28] Recent lone wolf attacks include the murder of Dutch politician Pim Fortuyn by animal rights activist Volkert van der Graaf in 2002, the Fort Hood assassinations by Nidal Hasan in 2009, the Toulouse and Montauban shootings by Mohammed Merah, targeting Muslim French soldiers and Jewish civilians in 2012, and the attacks carried out by the Ukrainian student Pavlo Lapshyn in 2013, who was convicted of murdering an elderly Muslim man and plotting explosions near mosques in England.[29] Doubtless the intense media coverage surrounding these events has made it seem as though such attacks have increased rapidly in recent years. A major research study by Christopher Hewitt reinforces this claim. Hewitt claims that lone wolf terrorism 'has greatly increased in recent decades'.[30] His research also demonstrates important geographical differences in the incidence of such terrorism. 'American terrorism', Hewitt argues, 'differs from terrorism in other countries in that a significant proportion of terrorist attacks have been carried out by unaffiliated individuals rather than by members of terrorist organisations'.[31] During the period 1955–77, 7% of all victims of terrorism in the US were reportedly killed by unaffiliated individuals, but during 1978–99 this percentage rose to 26. Hewitt's analysis, however, also includes attacks by couples and by three persons (which account for roughly 25% of cases), since his definition of lone wolf terrorism encompasses groups of up to four people.

According to a study by the Dutch Institute for Safety, Security and Crisis Management (COT), a total of 72 lone wolf terrorist incidents accounted for only 1.28% of the total number of terrorist incidents between 1 January 1968 and 1 May 2007 in the US, Germany, France, Spain, Italy, Canada and Australia.[32] Similarly Spaaij's study identified only 88 lone wolf cases across 15 countries, including the US, during 1968–2010. Lone wolf terrorism accounted for 1.8% of all terrorist incidents (198 out of a total of 11,235).[33] Although highly salient, acts of lone wolf terrorism appear to be rare. Like Hewitt's, however, Spaaij's study indicates that lone wolf terrorism has been significantly more prevalent in the US than in other Western countries, with the US cases (30 in total) accounting for almost 42% of all cases.[34] Furthermore, Spaaij's study reveals that between the 1970s and 2000s there was a marked increase in the number of lone wolf terrorist attacks not only within the US but also, more rapidly, in other countries. Spaaij's data show that the total number of lone wolf terrorist attacks per decade increased by 45% (from 22 to 32) in the US and by an extraordinary 412% (from eight to 41) in the other 14 countries combined. In total the number of lone wolf attacks rose from 30 in the 1970s to 73 in the 2000s, an increase of 143%.[35]

It has been suggested that lone wolf terrorist attacks are not generally very sophisticated, destructive or lethal occurrences,[36] and remain 'largely amateur' in their planning and execution when compared to acts committed by vast terrorist organisations, primarily as a result of limited resources, skills and capabilities.[37] Conversely other commentators have argued that in certain conditions the impact of lone wolf terrorism can be relatively large and unpredictable, and that, with access to high technologies, lone wolf terrorists may be 'very destructive'.[38] The 22 July 2011 attacks in Norway, which claimed a total of 77 lives, are an important reminder of this point. Moreover, lone wolf terrorists may be more difficult to detect and pose specific counterterrorism challenges. According to Bakker and de Graaf, 'lone operator terrorist acts should be considered "black swan" occurrences that are almost impossible to categorize or systematize, let alone forecast'.[39] In the next sections this article will focus on the challenges that the lone wolf attacks on 22 July 2011 presented for crisis management and contingency planning in Norway, and will explore possible counterterrorist approaches to the threat posed by lone wolf terrorism.

Lone wolf terrorism in Norway

The twin terror attacks on 22 July 2011 in Norway were completely unexpected and unprecedented. Norway is a small, multiparty parliamentary state with a strong democratic tradition and is often ranked among the nations with the highest levels of trust and civic engagement in the world.[40] It has, until 22 July 2011, been generally regarded as a 'peaceful, open and robust democracy',[41] with limited previous experience with terrorism and political violence.[42] The terror attacks on Oslo and Utøya were quickly characterised as the most destructive and devastating since the Second World War, and Norway's structures of crisis and contingencies management were put to the test. According to Boin and Lagadec, 'there is no such thing as a routine crisis'.[43] A crisis can be seen as an extreme situation that serves to disrupt core values or life-sustaining systems, and which requires a simultaneous coordinated response by many organisations.[44] Crises have been described as 'high impact, low probability' or 'black swan' events.[45] Decision-makers are thus confronted with hard dilemmas about non-routine situations that are characterised by severe threat, time pressure and high levels of uncertainty.[46] The terrorist attacks in Oslo and Utøya included all these elements.

On 22 July 2011 the first notification of a bomb explosion outside a main ministerial building in Oslo was registered by the police at 15:26.[47] The first police patrols arrived at the site a few minutes later, followed swiftly by the ambulance service. According to Christensen et al., the attack was quickly defined as a security crisis and one that 'demanded action from all levels and coordination across structures'.[48] The first emergency phone-calls reporting the attacks on the island of Utøya were registered at 17:25.[49] The local police district requested assistance from the national 'emergency unit' (*Beredskapstroppen*) or Delta force team, a national counterterrorism unit, at 17:38. The Delta force team (eight police officers in total) had already been deployed to Utøya from Oslo, but were held up in traffic. Further delays resulted from complications and miscommunication regarding meeting point and command. The first local police patrol arrived at Utvika pier point on the lake shore at 17:52 and the Delta force unit arrived 14 minutes later. Both teams joined forces and were transported to the island by boat. However, the engine failed after roughly 600 metres and the officers involved in the operation had to be rescued and taken to Utøya by two civilian craft. Moreover, there was no police helicopter available to provide air support or intervention. The first members of the Delta force team arrived on Utøya at 18:27, but were directed to the north of the island by a group of young people. The second group of officers arrived at 18:28 and proceeded to the south of the island after hearing gunshots. Eventually Anders Behring Breivik was apprehended without any resistance at 18:34.[50]

Christensen et al. reveal that 'a complex web of authorities' was responsible for emergency preparedness and crisis management on 22 July.[51] In Norway the Cabinet has an overall responsibility for security, and the Ministry of Justice normally takes on a lead coordinating role in a national crisis. The main operational units under the Ministry of Justice are the Directorate of the Police, the Police Security Service and the Joint Rescue Coordination Centres. The Directorate of the Police is responsible for the professional direction and follow-up of the police, and can assist the local chief of police in a crisis situation. The Police Security Service provides information and intelligence covering counterterrorism and counter-extremism, and warning and threat assessments, and provides personal protection. The Joint Rescue Coordination Centres are responsible for coordinating major search and rescue operations. At the local level local police districts are responsible for tactical decisions and operations. The national police force is divided into 27 police districts. Two different police districts were hit by the attacks of 22 July: Oslo and Nordre Buskerud. The Delta force team is organised within the police district of Oslo. All these institutions were operative during the attacks in July 2011.[52]

The Norwegian authorities' immediate response to the attacks, both strategically and operatively, was at first largely praised. Norway's then prime minister, Jens Stoltenberg, was heralded for responding to the espoused ideology of the accused perpetrator of the 22 July attacks by stating that 'the Norwegian response to violence is more democracy, more openness and greater political participation'.[53] He went on to comment that 'it's absolutely possible to have an open, democratic, inclusive society, and at the same time have security measures and not be naïve . . . I hope and also believe that the Norway we will see after [22 July] will be more open, a more tolerant society than what we had before'.[54] Doubtless the symbolic effect of this rhetoric and the initial speeches by the nation's leaders were significant and were pointedly different from declarations of war by the US president after 9/11.[55] Fimreite et al. point out that how authorities and particularly political leaders handle a crisis is essential for successful crisis management.[56] Three days after the attacks hundreds of thousands of Norwegians took part in what have been called 'the Rose marches'. All over the country flowers were laid down and candles were lit in a silent protest against the extremist views and acts of terror, and as a token of respect for the victims of the attacks. The statements and speeches of the nation's leaders, in particular the prime minister but also members of the royal family and other executive leaders, seemed to raise awareness in Norway that terror seeks to destroy trust, and apparently resulted in a mobilisation of core values in Norwegian society.[57]

However, over the following months, and as the crisis became more distant, the capacity and preparedness of the Norwegian authorities in crisis situations came to be heavily debated. Criticism was based on several internal evaluations, and on the report of a specially appointed independent commission investigating the incidents of 22 July: the 22 July Commission. The mandate of the 22 July Commission set out that it would examine the 'ability of the authorities and society to disclose plans of attacks and prevent attacks, protect themselves against the consequences of any future attacks, and deal with the situation during and after such events'.[58] The Commission published its 482-page report in August 2012, and was exceptionally critical on all levels: the prime minister was criticised for lacking emergency preparedness and sufficient awareness of risk; the government was accused of lacking implementation powers in the area of internal security and, simultaneously, doubts were raised as to whether the responsible authorities, especially the police, were at all prepared and adequately equipped for emergency preparedness and crisis management. Several key points were raised in relation to contingency planning and crisis management in Norway.

First and foremost the 22 July Commission argued that the terrorist attack that day could have been prevented through effective implementation of already adopted security measures.[59] The report was critical of what it identified as bureaucratic failings between departments that had left government buildings in the centre of Oslo vulnerable to attacks. The inquiry team raised questions as to why the street outside the government buildings – Grubbegata – was poorly protected and not closed to traffic, as had been recommended seven years previously. The Commission also revealed that established police protocol demanding immediate action in a crisis situation was not followed during the attacks. Although it was clear that a terrorist attack had been carried out, no immediate nationwide alert was given, no roadblocks or observation posts were set up, no attempt was made to mobilise helicopters, nor did the operation centre take up offers from neighbouring police districts. The operation centre was 'simply overloaded', the report concludes, to the extent that staff did little to prevent further attacks during 'the acute phase' of the terrorist attack.[60]

The report also identified two critical points that were linked to defective police work. First, information on Breivik's escape car was available but not followed up. The report gives details of a phone call made by a pedestrian 10 minutes after the Oslo bomb exploded, giving police a good description of a man carrying a pistol and wearing protective clothing. The person also gave the vehicle's licence plate number. According to the report, the operator passed the message on but the tip was not followed up properly.[61] Second, the police were criticised for their failure to use a helicopter once alerted to the shootings, and for the excessively long response time to reach the island of Utøya. Police were grappling with the wide damage inflicted by the bomb attack in the government quarter in central Oslo when word of the shootings on Utøya came.[62] The Delta force team then had to drive to the scene of the attack rather than take a helicopter because all police helicopter pilots were on vacation. The first boat they tried to take to the island broke down, forcing the emergency unit to take two civilian boats to the island instead. Critics have argued that the delayed response probably cost dozens of lives.[63]

In addition to criticism of the police, another issue was the coordination between different emergency units during the incident and the functionality of the emergency communication network. The report revealed that communication problems across and between emergency units 'caused resources to miss each other'.[64] Helicopters and the armed forces' other resources were requested 'so late' that they did not provide much help for the police operation.[65] In addition, the armed forces had no standby resources to support the police during an emergency. One explanation was that the armed forces' emergency helicopter services were being used for international operations in Afghanistan. However, the report revealed serious coordination problems between the different agencies involved, and recommended that the police airborne capacity was an important discussion point for future crisis preparedness.[66]

Another critical issue raised by the report was whether the Norwegian Police Security Service could have detected the activities of the perpetrator and taken action before the attacks. It argued that the Police Security Service in fact could have detected and prevented the terrorist attacks with better ways of working and a broader focus. Breivik might have been stopped seven months before the attacks, had Norway's internal intelligence service acted on a tip from customs officials, who flagged a suspicious purchase of potential bomb-making chemicals from Poland, but intelligence service failures meant that Breivik was not put on any watch list. A more effective response to warnings about Breivik would have revealed that he had been buying semi-automatic assault rifles and frequenting far-right websites. Furthermore, although Norway is a country with a large number of weapons, the Commission was also critical of 'inadequate' weapons controls and recommended stricter control of arms and chemicals.[67]

As noted by Christensen et al., the 22 July Commission report revealed 'a number of well-known problems within the policy field of crisis management and internal security',[68] concerning a lack of resources, ambiguous chains of responsibility and competences, a lack of coordination between relevant ministries and agencies, and internally within the police force. The criticism of the Norwegian authorities highlights the need for a swift law enforcement and emergency response to terrorist attacks that is well prepared, well coordinated, well equipped in terms of human resources and technical skills and capabilities, and repeatedly exercised.[69] Further, good teamwork between law enforcement agencies and other emergency services is clearly imperative during an emergency of the magnitude seen on 22 July.

The Commission concluded their report by stating that the Norwegian legislation in the area of civil protection had struck 'a reasonable balance between transparency and security', and reflected, in most areas, 'sensible ambitions for the level of security' in Norwegian society.[70] However, it added that the main challenge lay in engendering consistency between 'words and actions'.[71] In order to meet this challenge, the Commission suggested 31 different measures, including reforms within the government system and police, to legislation and within the armed forces, the emergency sector and the Police Security Service. These recommendations, however, do not represent radical changes from previous arrangements.[72] Nevertheless, the report launched an intense debate on the need to strengthen counterterrorist approaches in Norway. In May 2013 the government presented a proposal for legislative changes to handle 'solo-terrorism'.[73] The new proposals include the criminalisation of the planning of terrorist activities, receiving terrorist training and of active engagement in terror organisations. New investigation methods were also proposed and the maximum sentence for terrorism was proposed to increase from 21 to 30 years.[74] Furthermore, in the state budget for 2012 there was a significant increase in resources allocated for internal security and police operations.[75]

Organisational changes, however, have been minor. The name of the Ministry of Justice has been changed to the Ministry of Justice and Public Security, but the report evaluating the new structure of the ministry followed a very cautious approach and no major changes from previous arrangements have been suggested.[76] Even though new security measures will most probably be implemented in Norway, government authorities are well aware of the costs to democracy of such measures, and the infringement of civil liberties and due process rights. Soon after the attacks the then director of Norway's Police Security Service, Janne Kristiansen, stated that 'the trick is to make it easier to catch this kind of person, without taking away the openness and liberties more than necessary'.[77] Furthermore, the lone wolf operational model presents a number of critical challenges for crises and contingencies management. The next section aims to outline some of these key challenges and assess possible counterterrorist approaches to the threat posed by lone wolf terrorism.

Countering lone wolf terrorism

The primary challenge to preventing a lone wolf terrorist attack is that, by definition, lone wolves operate alone. As Fred Burton argues:

> When investigating a militant organization it is possible for law enforcement or intelligence agencies to plant informants within the group. Even small, insular groups are vulnerable because it is not uncommon for one or more members of the group to get cold feet and inform authorities about the group's plans to commit acts of violence. With a lone wolf, however, there is no such possibility of infiltration or betrayal. If the suspect never discusses his or her plans with anyone else, he or she can easily fly under law enforcement radar. In most cases, these kinds of individuals can be highly successful in carrying out an attack, especially against vulnerable soft targets ... Mentally disturbed lone wolves pose particular problems because they often have an extremely narrow focus of interest and cannot be diverted to an easier target by heightened security measures...Mentally disturbed lone wolves also frequently have an almost total disregard for the consequences of their actions, and quite often show no concern about escaping after they attack.[78]

Indeed, it is extremely difficult to detect terrorists operating individually because they are likely to be relatively socially isolated and may avoid contact with others.[79] Lone wolves also create difficulties because they display a wide variety of backgrounds with a wide spectrum of ideologies and motivations, which make it hard to predict from which environment they might stem.[80] It is also extremely difficult to differentiate between those extremists who intend to commit attacks and those who simply express radical beliefs, or hollow threats. In many countries freedom of speech is a fundamental freedom which limits the possibilities of investigating non-violent radical scenes. As Bakker and de Graaf point out, 'knowing that all terrorists are radical but that most radicals are not terrorists, it is extremely difficult to single out potential lone wolves before they strike, even with the help of the most sophisticated intelligence gathering tools'.[81] Furthermore, lone wolves usually carry out a single attack, rather than a more prolonged terrorist campaign characteristic of most conventional group-actor terrorism. It is within this context that Europol has stated that the activities of lone wolf terrorists 'can be unpredictable and difficult to prevent'.[82] Bakker and de Graaf go even further by contending that lone operator terrorists 'provide the most puzzling and unpredictable form of terrorism' for counterterrorism organisations, the police and intelligence communities.[83]

Notwithstanding the challenges presented by lone wolf operatives, it is possible to identify some commonalities among government and community responses to the threat of such terrorism. The most dominant response to acts of lone wolf terrorism adopted by authorities is to implement repressive strategies, such as criminal investigation, increased security and heightened surveillance.[84] Law enforcement agencies play a pivotal role in these counterterrorist operations, but the pros and cons of repressive approaches should be considered carefully. Security measures can doubtless hinder the operation of a lone wolf terrorist, and may even discourage copycat behaviour or deter less determined and less capable terrorists.[85] In the aftermath of the 22 July Commission report the Norwegian prime minister emphasised that increased security was to be attained through more controls, visible police, adequate emergency organisation, exercises and training, and the right equipment.[86] Enhanced security may thus benefit a nation by lessening the vulnerability of critical infrastructure and key resources, essential leadership and major events to lone wolf terrorist attacks. However, the effectiveness of such measures remains under-researched, and should not be overstated. As Spaaij notes, increasing security in one area may have the unintended consequence of encouraging lone wolf terrorists to attack other 'soft' targets, or devise tactics to circumvent such measures.[87] Furthermore, a number of authors has raised the important question, 'how much is too much security?'. As Zedner argues, 'whilst security promises to enhance subjective feelings of safety its pursuit often entails increased insecurity', and in fact erodes civil liberties and human rights.[88]

The collection and analysis of intelligence in order to prevent acts of lone wolf terrorism is another major strand of counterterrorist responses by law enforcement agencies. In a context where anticipation and predictability are low, intelligence-led operations must focus not on knowing *who* is planning an attack, but rather on knowing *how* such attacks are formulated.[89] Thus, in the aftermath of the 22 July 2011 attacks, the question emerged as to what the authorities could have done to identify or prevent Breivik's purchases of bomb-making equipment and specialist weapons, as well as other suspicious behaviour. Indeed, lone wolf terrorist attacks are usually preceded by a careful and prolonged preparatory phase.[90] Lone wolves are thus constrained by the terrorist attack cycle, and are vulnerable to detection at several different stages, most notably during the pre-attack stage when weapons are procured and targets are selected.[91] At the pre-attack stage intelligence services can pick up key signals in relation to, for example, the purchase of chemicals for improvised explosive devices or the acquisition of weapons or suspicious internet activity.[92] This information can be fused and can shed light on how lone wolf attacks are assembled, increasing the ability of interdiction forces to operate successfully to the 'left of the bang', or before the attack.[93] Communication between the police and local communities is often critical for law enforcement agencies and intelligence analysts at this stage. Spaaij's study reveals that there have been several occasions where police and intelligence agencies were only alerted to an alleged plot following reports from shop owners or internet companies over suspicious purchases.[94]

Nevertheless, whether these processes open up possible avenues for the detection and prevention of a lone wolf terrorist act depends on both the quality of the data gathered, and its subsequent analysis and follow-up. Furthermore, intelligence-led responses may also sharply infringe on the privacy and civil liberties of individuals. This type of information-gathering requires the police and security services to use more intrusive procedures, such as undercover operations, electronic surveillance and the monitoring of internet and social networking activity.[95] As Spaaij points out, intelligence-led responses to lone wolf terrorist attacks are not only hampered by the fact that it is difficult to distinguish between those extremists who intend to commit attacks and those who simply express radical beliefs, but also raise important questions concerned with the erosion of human rights, the infringement of liberties, and the prevention of intelligence excesses.[96]

Importantly lone wolf terrorism does not take place in a social vacuum. As discussed earlier, lone wolves often exhibit a degree of commitment to, and identification with, extremist movements and communities of support. In the light of this, counter-narratives are also an important component of a comprehensive response strategy aimed at counter-radicalisation. Shone contends that security agencies must work collaboratively with religious and community leaders to ensure that community alienation is reduced and non-radical ideologies are propagated, thus delegitimising extremist beliefs.[97] Previous research has shown that good relationships with the public and local community leaders can create and maintain credibility, foster high levels of trust and prevent disputes from escalating into crises.[98] Other studies emphasise the importance of stimulating democratic principles and the openness and inclusiveness of society, limiting isolation and opening democratic paths and social opportunities.[99] The radicalisation process for lone wolf terrorists, however, remains poorly understood. The underlying mechanisms and processes that lead future lone wolves to adopt increasingly extreme political, religious or social ideals and aspirations and reject social norms demand scientific research.[100] In conjunction with repressive responses, then, the value of conciliatory or democratic responses of this nature should not be underestimated.

Lone wolf terrorist attacks cannot always be prevented, however, and can therefore create painful and costly ramifications for both victims and society. Thus it is important for governments to strengthen emergency preparedness and resilience to anticipate and minimise the impact of an attack. In the context of lone wolf terrorism the concept of resilience is of major import, and one that has been widely acknowledged within the counter-terrorism and disaster management literature.[101] The definition of resilience is somewhat contested but generally includes 'activities designed to reduce or eliminate risks to persons or property or to lessen the actual or potential effects or consequences of an incident ... [that] may be implemented prior to, during, or after an incident'.[102] As shown above, in the aftermath of the 22 July 2011 attacks in Norway the police and security services came under scrutiny for the slow pace of their response to the incidents, the lack of clarity in command and the seemingly poor coordination within and across emergency services. Such criticism highlights the need to strengthen planning processes and create generic responses for events that can be rare and unexpected. A broad strategy for countering lone wolf terrorism would thus encompass a mix of interdiction, intelligence gathering, counter-radicalisation, emergency preparedness and resilience, within a framework that respects human rights and prevents the erosion of civil liberties as much as possible.

Conclusion

Lone wolf terrorism is not a new phenomenon and has been traced back at least to 19th century anarchism. Available studies of terrorist attacks by individual perpetrators suffer from a lack of consensus regarding the definition of lone wolf terrorism. Even though studies have produced well founded knowledge about solo-actor terrorism, serious questions regarding the validity of findings and their generalisation remain. Furthermore, while lone wolf terrorism has recently been presented as a growing threat across Western nations, existing research indicates that lone operator terrorism is a marginal phenomenon and less dangerous than conventional group-actor terrorism. However, Breivik's massacre of 77 people on 22 July 2011 in Norway demonstrates the challenges that such terrorism can present for crisis management and contingency planning in the twenty-first century, and the Norwegian authorities have come under serious scrutiny for lacking capacity and preparedness in crisis situations.

While the Norwegian authorities understandably intend to make their surveillance and security methods more effective and their critical infrastructure more robust, the predominant response in Norway to the attacks has been defined by a 'belief that there are other values besides security that matter a great deal and that pursuing security above all other values, in a quest for absolute safety, is both self-destructive and futile'.[103] As Jens Stoltenberg, whose office was the target of the bomb and whose Labour Party was the sponsor of the camp on the island of Utøya, stated in response to Breivik's attacks, 'We must – and will – meet terror with more democracy, not less. We must not lock up Norwegian society. That would be piling tragedy upon tragedy'.[104] Indeed, Stoltenberg strongly defended the right to speak freely, even if it includes extremist views such as Breivik's.

Central to this response, then, is learning to live with risk.[105] According to Furedi, this type of response is the most effective way of countering and reducing the threat of lone wolf terrorism. Furedi states that 'the way that society responds to acts of individual terror represents a far greater threat than the destruction caused by the lone wolf himself', and that, while open societies are susceptible to discrete acts of terrorism, terrorist acts 'cannot seriously threaten the integrity of society nor undermine the way of life of a nation'.[106] It could be argued therefore that the most effective and principled response to future acts of terrorism in Norway will involve refusing to sacrifice liberty and transparency and maintaining the openness, trust and democratic values that distinguish Norwegian government and society.

Acknowledgments

I am grateful to Clive Walker for inspiring this research project, for helpful discussions during the research process, and for his comments on an earlier draft of this paper. Any mistakes that remain are, of course, mine.

Notes

1. See http://www.domstol.no/upload/OBYR/Internett/Nyheter/Terrorsaken/toslo-2011-188627-24.pdf; http://www.domstol.no/no/Enkelt-domstol/22-7/Nyheter-om-227-saken/Breivik-convicted-to-permanent-detention/, August 24, 2012. See also BBC News 'Anders Behring Breivik: Norway Court Finds Him Sane', August 24, 2012, http://www.bbc.co.uk/news/world-europe-19365616.
2. Aamer Madhan, 'Obama: "Lone Wolf" Attack is Biggest Concern', *National Journal*, August 17, 2011, http://www.nationaljournal.com/whitehouse/obama-lone-wolf-attack-is-biggest-concern-20110816.
3. 'Lone Wolf Types Emerge as Terror Threat: ASIO', *Canberra Times*, September 11, 2011, http://www.canberratimes.com.au/national/lone-wolf-types-emerge-as-terror-threat-asio-20110919-1wqaz.html.
4. Europol, *Annual Terrorism Situation and Trend Report* (The Hague: Europol, 2012), 9, https://www.europol.europa.eu/sites/default/files/publications/europoltsat.pdf. See also EU Counter-terrorism Coordinator, *Preventing Lone Actor Terrorism – Food for Thought* (Brussels, 2012).
5. One reason for this is that terrorism has generally been understood to be a group-based activity endorsed by a wider organisation, with little room for individually planned and undertaken violence.
6. See, for example, Christopher Hewitt, *Understanding Terrorism in America: From the Klan to al-Qaeda* (New York: Routledge, 2003); Instituut voor Veiligheids- en Crisismanagement (COT), *Lone-Wolf Terrorism* (The Hague: COT, 2007); Ramon Spaaij, 'The Enigma of Lone Wolf Terrorism: An Assessment', *Studies in Conflict Terrorism* 33 (2010): 854; Spaaij, *Understanding Lone Wolf Terrorism: Global Patterns, Motivations and Prevention* (New York: Springer, 2012); Raffaello Pantucci, 'What Have We Learned about Lone Wolves from Anders Behring Breivik?', *Perspectives on Terrorism* 5, nos. 5–6 (2011): 27; Randy Borum, 'Informing Lone Investigations', *Criminology and Public Policy* 12, no. 1 (2013): 103; George Michael, *Lone Wolf Terrorism and the Rise of Leaderless Resistance* (Nashville, TN: Vanderbilt University Press, 2013); and Jeffrey D. Simon, *Lone Wolf Terrorism: Understanding the Growing Threat* (New York: Prometheus Books, 2013).
7. Simon, *Lone Wolf Terrorism*, 37–8.
8. Hewitt, *Understanding Terrorism in America.*
9. See, esp., Matthew Feldman, 'Comparative Lone Wolf Terrorism: Toward a Heuristic Definition', *Democracy and Security* 9 (2013): 270.
10. Fred Burton and Scott Stewart, 'The "Lone Wolf" Disconnect', *STRATFOR: Global Intelligence*, January 30, 2008, http://www.stratfor.com/weekly/lone_wolf_disconnect.
11. Ibid.
12. Recent evidence submitted to the UK-based Intelligence and Security Committee of Parliament (ISC) by Andrew Parker, Director General of the Security Service, suggests that the rise of the internet has resulted in 'a small rise in the number of cases we see of individuals who become radicalised, and sometimes [turn] to violent action'. ISC, 'Evidence given by Sir Iain Lobban, Mr Andrew Parker, Sir John Sawyers' (uncorrected transcript, 2013), 11.

13. See Spaaij, 'The Enigma of Lone Wolf Terrorism'; and Spaaij, *Understanding Lone Wolf Terrorism*. See also COT, *Lone-Wolf Terrorism*.
14. Spaaij, 'The Enigma of Lone Wolf Terrorism', 856.
15. Raffaello Pantucci, *A Typology of Lone Wolves: Preliminary Analysis of Lone Islamist Terrorists* (London: International Centre for the Study of Radicalisation and Political Violence, 2011), 19.
16. Gerry Gable and Paul Jackson, *Lone Wolves: Myth or Reality?* (Ilford: Searchlight Report, 2011), 7.
17. Ibid., 82.
18. Ibid.
19. For example, Breivik claimed in his manifesto to have started working on manufacturing explosives by searching the internet. Anders Breivik, *2083: A European Declaration of Independence*, 2011, http://www.washingtonpost.com/r/2010-2019/WashingtonPost/2011/07/24/National-Politics/Graphics/2083+-+A+European+Declaration+of+Independence.pdf.
20. Feldman, 'Comparative Lone Wolf Terrorism', 280.
21. See Boaz Ganor, 'Defining Terrorism: Is One Man's Terrorist Another Man's Freedom Fighter?', *Police Practice and Research* 3, no. 4 (2002): 287.
22. Walter Laqueur, *The New Terrorism: Fanaticism and the Arms of Mass Destruction* (New York: Oxford University Press, 2000).
23. See Jeffrey J. Kaplan, 'Leaderless Resistance', *Terrorism and Political Violence* 9, no. 3 (1997): 80.
24. Ibid.
25. Louis Beam, 'Leaderless Resistance', *The Seditionist* 12, (1992): 5 (first published 1983), http://www.louisbeam.com/leaderless.htm.
26. See Metzger, cited in Petter Nesser, 'Research Note: Single Actor Terrorism – Scope, Characteristics and Explanations', *Perspectives on Terrorism* 6, no. 6 (2012): 62.
27. Breivik, *2083*.
28. See, for example, Matthew Goodwin, '"Lone wolves" such as Pavlo Lapshyn are part of a bigger threat', *Guardian*, October 23, 2013, http://www.theguardian.com/commentisfree/2013/oct/23/pavlo-lapshyn-extremist-bomber-lone-wolves.
29. See http://www.cps.gov.uk/news/latest_news/man_jailed_for_mosque_explosions_and_murder_of_elderly_muslim_man/. Furthermore, recent statements from al-Qaeda underline the militant Islamist groups' increased strategic shift towards the practice of so-called 'individual jihad' or solo terrorism, encouraging individuals to execute small-scale, less technically complex attacks in Western countries without the active assistance of any larger organisation. See Beau D. Barnes, 'Confronting the One-man Wolf Pack: Adapting Law Enforcement and Prosecution Responses to the Threat of Lone Wolf Terrorism', *Boston University Law Review* 92 (2012): 1613; and Europol, *TE-SAT 2012: EU Terrorism Situation and Trend Report* (The Hague: Europol, 2012): 17.
30. Hewitt, *Understanding Terrorism in America*, 79.
31. Ibid., 78.
32. COT, *Lone-Wolf Terrorism*, 16–17.
33. Spaaij, *Understanding Lone Wolf Terrorism*, 27.
34. This can, in part, be explained by the relative popularity of the leaderless resistance concept among right-wing militia and anti-abortion activists in the US. Ibid., 31.
35. Ibid., 32. Importantly, Spaaij also acknowledges the limitations of his dataset, such as the likelihood of a 'hidden figure' of lone wolf terrorists, particularly in historical periods and regions of the world with low media coverage. Nesser also points out that, while Spaiij's lone wolf study is an important contribution, the dataset referred to in the study almost exclusively registers successful attacks. Nesser argues that failed attacks should also be taken into consideration, given that terrorist plots often fail or are thwarted. Nesser, 'Research Note'.
36. According to Spaaij, *Understanding Lone Wolf Terrorism*, 92, the lethality rate of lone wolf terrorist attacks averages around 0.62 fatalities per incident, compared to about 1.6 deaths per attack in other terrorist attacks. Furthermore, there is no evidence to suggest that the overall lethality of lone wolf terrorism is on the increase compared with the growing lethality of terrorism in general. Walter Enders and Todd Sandler, 'After 9/11: Is it All Different Now?', *Journal of Conflict Resolution* 49, no. 2 (2005): 259.
37. Europol, *TE-SAT 2012*, 18.

38. V.I. Vasilenko, cited in Spaaij, *Understanding Lone Wolf Terrorism*, 28. See also David Anderson, *The Terrorist Acts in 2012: Report of the Independent Reviewer on the Operation of the Terrorism Act 2000 and Part I of the Terrorism Act 2006* (London: The Stationery Office, 2013), paras 2.21–2.22.
39. Edwin Bakker and Beatrice de Graaf, 'Preventing Lone Wolf Terrorism: Some CT Approaches Addressed', *Perspective on Terrorism* 5, nos. 5–6 (2003): 43.
40. Dag Wollebæk et al., 'After Oslo and Utøya: How a High-trust Society Reacts to Terror', *Political Science and Politics* 45, no. 1 (2012): 32.
41. Tom Christensen, Per Lægreid, and Lise Hellebø Rykkja, 'After a Terrorist Attack: Challenges for Political and Administrative Leadership in Norway', *Journal of Contingencies and Crisis Management* 21, no. 3 (2013): 167.
42. According to Simon, *Lone Wolf Terrorism*, 49, between 1970 and 2010 there were 15 terrorist attacks in Norway, with one person killed and 13 wounded in those incidents.
43. Arjen Boin and Patrick Lagdec, 'Preparing for the Future: Critical Challenges in Crisis Management', *Journal of Contingencies and Crisis Management* 8, no. 4 (2000): 185.
44. Arjen Boin, cited in Christensen et al., *After a Terrorist Attack*, 168.
45. See Bernice Lee and Felix Preston with Gemma Green, *Preparing for High Impact, Low Probability Events* (London: Chatham House, 2012).
46. Paul t'Hart, Uriel Rosenthal, and Alexander Kouzmin, 'Crisis Decision Making: The Centralization Thesis Revisited', *Administration and Society* 25, no. 1 (1993): 12.
47. Norges offentlige utredninger (NOU), Official Norwegian Reports, *Rapport fra 22. juli-kommisjonen* (Oslo: Departementenes servicesenter Informasjonsforvaltning, Government Administration Services, Information Management, 2012).
48. Christensen et al., 'After a Terrorist Attack', 12.
49. NOU, *Rapport fra 22. Juli-kommisjonen*.
50. Ibid. Before his arrest Breivik contacted the police on two occasions during his attack (recorded at 17:59 and 18:24), presenting himself and wanting to surrender, but both of his calls were cut short as a result of severe capacity problems within the emergency telephone system. Ibid.
51. Tom Christensen, Per Lægreid, and Lisa Rykkja, *How to Cope with a Terrorist Attack? A Challenge for the Political and Administrative Leadership* (COCOPS, 2012): 11, http://www.cocops.eu/wp-content/uploads/2012/08/COCOPS_workingpaper_No6.pdf.
52. Ibid.
53. Helen Pidd and James Meikle, 'Norway will Not be Intimidated by Terror Attacks, Vows Prime Minister', *Guardian*, July 27, 2011, http://www.theguardian.com/world/2011/jul/27/norway-terror-attacks-prime-minister. See also Richard Orange, 'Jens Stoltenberg Interview: Prime Minister at the Helm during Breivik Shooting', *Daily Telegraph*, July 21 2012, http://www.telegraph.co.uk/news/worldnews/europe/norway/9418144/Jens-Stoltenberg-interview-prime-minister-at-the-helm-during-Breivik-shooting.html.
54. Michael Schwirtz, 'Norway's Premier Vows to Keep an Open Society', *New York Times*, July 27, 2011, http://www.nytimes.com/2011/07/28/world/europe/28norway.html?_r=0.
55. In his speech to the Congress in September 2001 President Bush's message was: 'Our war on terror begins with al-Qaeda but it does not end there. It will not end until every terrorist group of global reach has been found, stopped and defeated.' Cited in BBC News, 'Declining Use of "War on Terror"', April 17, 2007, http://news.bbc.co.uk/2/hi/uk_news/politics/6562709.stm.
56. Anne Lise Fimreite et al., 'After Oslo and Utøya: A Shift in the Balance between Security and Liberty in Norway?', *Studies in Conflict and Terrorism* 36 (2013): 839.
57. Wollebæk et al., 'After Oslo and Utøya', 35.
58. NATO, '22 July Commission's Report', August 14, 2012, http://www.norway-nato.org/eng/News/22-July-Commissions-report/.
59. NOU, *Rapport fra 22. Juli-kommisjonen*.
60. Ibid. See also BBC News, 'Norway Police "Could have Stopped Breivik Sooner"', August 13, 2012, http://www.bbc.co.uk/news/world-europe-19241327.
61. NOU, *Rapport fra 22. Juli-kommisjonen*.
62. See Ian MacDougall, 'Norway PM: Attacks Response to be "More Democracy"', July 27, 2011, http://news.yahoo.com/norway-pm-attacks-response-more-democracy-121223136.html.
63. See Michael Schwirtz, 'Norway's Premier Manages to Keep an Open Society', *New York Times*, July 27, 2011, http://www.nytimes.com/2011/07/28/world/europe/28norway.html?_r=0.

64. NOU, *Rapport fra 22. Juli-kommisjonen.*
65. Ibid.
66. See Christensen et al., *How to Cope with a Terrorist Attack?.*
67. NOU, *Rapport fra 22. Juli-kommisjonen.*
68. Christensen et al., *How to Cope with a Terrorist Attack?*, 20.
69. Spaaij, *Understanding Lone Wolf Terrorism.*
70. NOU, *Rapport fra 22. Juli-kommisjonen.*
71. Ibid.
72. See Christensen et al., *How to Cope with a Terrorist Attack?.*
73. See Fimreite et al., 'After Oslo and Utøya', 851.
74. Prop.131 L (2012–2013) Proposisjon til Stortinget (forslag til lovvedtak) 2005 mv. (forberedelse av terror m.m.). These proposals were approved by the Norwegian Parliament in June 2013. See Lovvedtak 104 (2012–1013) vedtak til lov om endringer i straffeloven 1902 og straffeloven 2005 mv. (forberedelse av terror m.m.). However, at the time of writing, they are yet to be implemented. The proposal to increase the maximum sentence from 21 to 30 years for acts of terror was, in fact, passed by the Norwegian parliament on 20 May 2005. Prop. 131 L (2012–2013) suggests that the implementation of a 30-year tariff for the most serious crimes, including acts of terror, should be pushed forward.
75. See Royal Ministry of Finance, *The National Budget 2012: A Summary* (Oslo, 2012), http://www.statsbudsjettet.no/Upload/Statsbudsjett_2012/dokumenter/pdf/nb_summary.pdf.
76. See Per Lægrid and Lisa H. Rykkja, *Coordination Practice: Coordinating for Internal Security and Safety in Norway* (COCOPS, 2013), http://www.cocops.eu/wp-content/uploads/2013/06/Norway_CGov_Internal-Security.pdf; and Christensen et al., *How to Cope with a Terrorist Attack?.*
77. Adam Geller and Ian MacDougall, 'Massacre Changes Norway's Thinking on Personal Liberty', *Washington Times*, August 7, 2011, http://www.washingtontimes.com/news/2011/aug/7/massacre-changes-norways-thinking-on-personal-libe/?page=all.
78. Fred Burton, 'The Challenge of the Lone Wolf', *Security Weekly*, May 30, 2007, http://www.stratfor.com/challenge_lone_wolf.
79. Breivik was clearly aware of this aspect of lone terrorist acts. In his manifesto he warned other potential terrorists that their chances of being apprehended would increase by 100% for every other person they involved in their plans. He stated: 'Don't trust anyone unless you absolutely need to (which should never be the case). Do absolutely everything yourself'. Breivik, *2083*, 844.
80. See Bakker and de Graaf, 'Preventing Lone Wolf Terrorism'.
81. Edwin Bakker and Beatrice de Graaf, 'Lone Wolves: How to Prevent this Phenomenon?', (The Hague: International Centre for Counter-Terrorism, 2010): 5, http://www.icct.nl/download/file/ICCT-Bakker-deGraaf-EM-Paper-Lone-Wolves.pdf.
82. Europol, *Annual Terrorism Situation and Trend Report* (The Hague: Europol, 2011), 15.
83. Bakker and de Graaf, 'Preventing Lone Wolf Terrorism', 46.
84. As Greenwald states, 'Over the last decade, virtually every terrorist plot aimed at the US – whether successful or failed – has provoked greater security and surveillance measures'. Glenn Greenwald, 'An Un-American response to the Oslo Attack', July 28, 2011, http://www.salon.com/2011/07/28/norway_4/print.
85. See Gus Martin, *Understanding Terrorism* (London: Sage Publications, 2003).
86. Christensen et al., *How to Cope with a Terrorist Attack?.*
87. Spaaij, *Understanding Lone Wolf Terrorism*, 101.
88. Lucia Zedner, 'Too Much Security?', *International Journal of the Sociology of Law* 31 (2003):163.
89. Alex Shone, 'Countering Lone Wolf Terrorism: Sustaining the CONTEST Vision', May 17, 2010, http://www.henryjacksonsociety.org/stories.asp?id=1582.
90. Breivik, for example, claims to have spent 'a total of nine years', planning his attacks. Breivik, *2083*, 8.
91. Shone, 'Countering Lone Wolf Terrorism'.
92. Breivik's manifesto revealed that he felt vulnerable to detection before his attacks. For instance, he created an extensive cover story that included renting a farm in order to explain his purchase of a large quantity of ammonium nitrate fertiliser. Breivik, *2083.*
93. Shone, 'Countering Lone Wolf Terrorism'.
94. Spaaij, *Understanding Lone Wolf Terrorism.*

95. See Gabriel Weimann, 'Lone Wolves in Cyberspace', *Journal of Terrorism Research*, 3, no. 2 (2012): 75.
96. Spaaij, *Understanding Lone Wolf Terrorism*.
97. Shone, 'Countering Lone Wolf Terrorism'.
98. Paul K. Davis and Kim Cragin, *Social Science for Counter Terrorism: Putting the Pieces Together* (Santa Monica, CA: RAND, 2009).
99. Froukje Demant et al., cited in Spaaij, *Understanding Lone Wolf Terrorism*, 90.
100. See Noémie Bouhana and Per-Olof Wikström, *Al Qa'ida Influenced Radicalisation*, Occasional Paper 97 (London: Home Office, 2011); and Tony Munton et al., *Understanding Vulnerability and Resistance in Individuals to the Influence of Al Qa'ida Violent Extremism*, Occasional Paper 100 (London: Home Office, 2011).
101. See, for example, Stephen Flynn, *The Edge of Disaster: Rebuilding a Resilient Nation* (New York: Random House, 2007); and Yossi Sheffi, *The Resilient Enterprise: Overcoming Vulnerability for Competitive Advantage* (Cambridge, MA: 2007).
102. Department of Homeland Security, *National Response Plan*, cited by Brian A. Jackson, *Marrying Prevention and Resiliency*, Occasional Paper (Santa Monica, CA: RAND, 2008), 3.
103. Greenwald, 'An Un-American Response to the Oslo Attack'.
104. 'Resilient and peaceful Norwegians won't allow attack to change their way of life', *The Mirror*, July 24, 2011, http://www.mirror.co.uk/news/uk-news/resilient-and-peaceful-norwegians-wont-allow-143643.
105. Since the attacks Stoltenberg and members of Norway's royal family have underlined the country's openness by making public appearances with little visible security.
106. Frank Furedi, 'Who's Afraid of the Big Bad "Lone Wolf"?', February 22, 2012, http://frankfuredi.com/index.php/site/article/537/.

Managing 'civil contingencies' in Australia

Michael Eburn

College of Law, Australian National University, Canberra, Australia

This paper identifies how 'civil contingencies' are addressed in Australia. It describes how responsibilities for emergency management are allocated across Australia's federated system of government and provides a critical review of the institutional and structural arrangements. It is argued that, even though Australia's emergency management arrangements have generally served the community well, there are undoubted limitations. One limitation is the lack of established Commonwealth response powers and weak arrangements to coordinate the Commonwealth response to emergency. It is further argued that Australia's arrangements focus on governments and their response to emergencies. They have, to date, largely omitted other players, in particular non-government actors, whether that is private industry, communities or individuals. There is evidence that this is changing with the recent overarching policy statement – the National Strategy for Disaster Resilience.

Introduction

This paper identifies how 'civil contingencies' are addressed in Australia. It describes how responsibilities for emergency management are allocated across Australia's federated system of government and provides a critical review of the institutional and structural arrangements that are in place. It is argued that, even though Australia's emergency management arrangements have generally served the community well, there are undoubted limitations. One limitation is the lack of established Commonwealth response powers and weak arrangements to coordinate the Commonwealth response to emergency. This has not proved problematic to date, but preparation for a catastrophic event, of national significance, should not be avoided simply because it has not yet happened. It is further argued that Australia's arrangements focus on governments and their response to emergencies. They have, to date, largely omitted other players, in particular non-government actors, whether that is private industry, communities or individuals. There is evidence that this is changing with the recent overarching policy statement – the *National Strategy for Disaster Resilience*. What is required now is work at the local level to develop true local resilience to the impact of inevitable natural hazards such as fire, flood and storm.

Governance arrangements

It may be thought that there are three tiers of government in Australia: federal, state and local as shown in Figure 1.

This model is, however, an oversimplification of Australia's federal arrangements. The Commonwealth (i.e. federal) government is a government of limited power. It is limited to making laws and otherwise dealing with matters allocated to it by the Commonwealth Constitution.[1] Notwithstanding the fact that the areas of Commonwealth interest are limited, the Commonwealth exercises significant power over all areas of Australian life, given its role as the national tax collector. Further, the High Court of Australia has held that the Commonwealth can use its broad powers to encroach into areas, traditionally believed to belong to the state, provided that the law in question can be described as a law 'with respect to' one of the listed areas of Commonwealth power. For example, reliance on the power to make laws with respect to external affairs allowed the Commonwealth to prohibit the construction of a dam within Tasmania for the purposes of generating electricity there,[2] and reliance on the power to make laws with respect to corporations allowed the Commonwealth to introduce significant reforms in the area of workplace agreements and employer–employee relations.[3]

The Australian states were the former colonies created by the UK following the arrival of the First Fleet in 1788. State governments have plenary powers to make law for peace, order and good government within the state. Under the Constitution some areas of law fall exclusively under the control of the Commonwealth but, in other areas, both the Commonwealth and the states may regulate. In the event of any inconsistency between a valid Commonwealth and a valid state law, the Commonwealth law prevails.[4] Subjects that are outside the permitted scope of Commonwealth interest remain an area of exclusive state regulation.

There are two mainland territories: the Australian Capital Territory and the Northern Territory. The Commonwealth parliament has the power to make laws with respect to territories,[5] but, rather than manage them directly, the Commonwealth has established institutions to allow self-government.[6] Further reference in this paper to 'the states' or 'a state' should be taken to include the self-governing territories. Although the territories may be subject to direct Commonwealth control, in practice they operate on virtually the

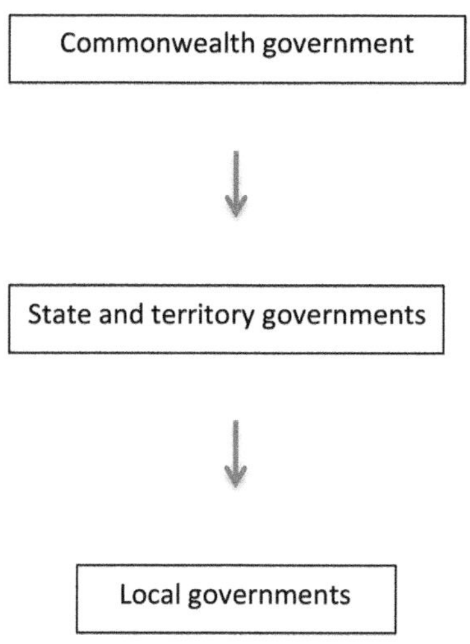

Figure 1. Three tiers of government.

same constitutional basis as the states. There was one notable exception when the Northern Territory's voluntary euthanasia law, the Rights of the Terminally Ill Act (NT), was directly disallowed by the Commonwealth Euthanasia Laws Act 1997 (Cth). Such direct intervention is unlikely today; for example, in 2013 the Australian Capital Territory introduced laws to allow same-sex marriage, contrary to Commonwealth government policy and the definition of marriage in the Marriage Act 1961 (Cth). The Commonwealth challenged that law in the High Court of Australia, rather than overruling it by direct intervention.[7]

Local governments operate in each jurisdiction except the Australian Capital Territory. Local governments are the product of state legislation and operate subject to state policy directives.[8] There are 565 local government authorities across Australia.[9]

A more accurate, and necessarily complex, diagram of the federal arrangements is shown below (Figure 2). Local government is clearly subject to state government and may be said to sit 'below' the states. Equally the territories are subject to direct Commonwealth regulation and are therefore subordinate to the Commonwealth (even though, as noted above, the Commonwealth is unlikely to directly intervene in territory governance). Where they are able to regulate on the same subject, then Commonwealth law does prevail, but the states have their own areas of exclusive law-making powers and in those areas they are not subject, or subordinate, to the federal government.

In this system the development and implementation of national policy requires cooperative government arrangements rather than imposition of policy from the Commonwealth government to subordinate state and local governments. The peak body for inter-government cooperation is the Council of Australian Governments (COAG). The members of

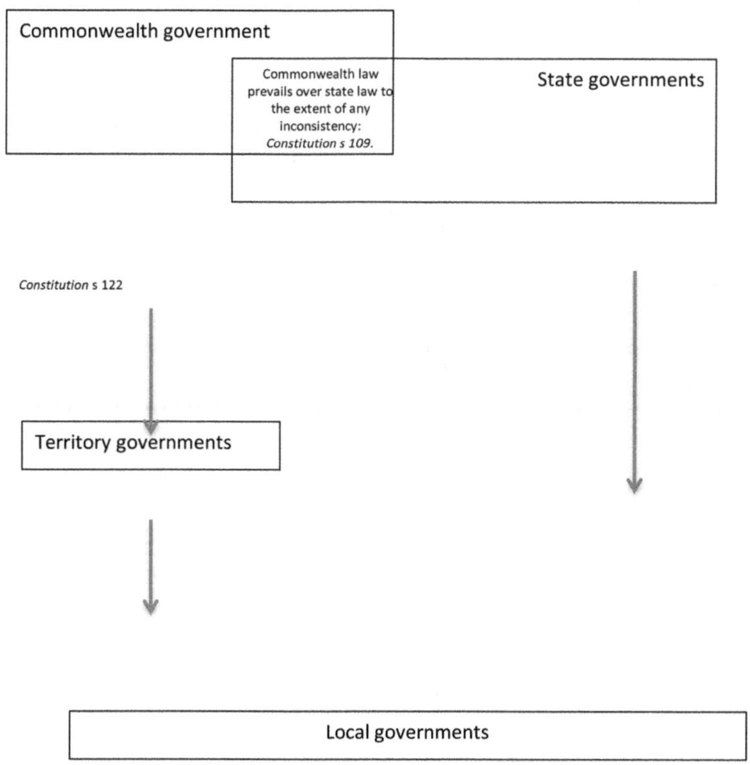

Figure 2. Schema of Australian government arrangements.

COAG are the prime minister, the premier or chief minister from each state and territory and a representative from the Australian Local Government Association.[10]

Responsibilities for emergencies or civil contingencies

The Commonwealth

The Commonwealth has a limited role in natural disaster planning and response. Issues of disaster and emergency management are not mentioned in the Australian Constitution, with the consequence that these matters are primarily matters for the states. The Commonwealth can make laws that have an impact upon all aspects of emergency management if those laws are laws 'with respect to' an area of interest to the Commonwealth. For example, the Commonwealth has the power to make laws with respect to 'telegraphic, telephonic, and other like services', 'meteorological observations', 'Insurance' and 'External Affairs'.[11] All these topics are important in preparing for and responding to an emergency. Further, the Commonwealth and its agencies are likely to be affected by any civil contingency and so the Commonwealth needs to plan to ensure business continuity – that is, how it and its agencies will deal with the impact of a disaster using their own capacity and how they will deal with extra demand for Commonwealth services following a disaster.

The most significant contribution from the Commonwealth is funding for projects designed to increase resilience to natural hazards, and funding recovery grants to the states and to individuals.[12] During the response phase to any disaster the Commonwealth can make federal assets, including the resources of the Australian Defence Force, available to the affected states. Commonwealth assistance is dependent upon the state's making a request and the request being approved by the Commonwealth Attorney General and relevant minister.[13]

Notwithstanding the Commonwealth's interest and direct involvement in disaster recovery, there is no Commonwealth counter-disaster or emergency management legislation. The failure of the Commonwealth to legislate in this area has, to date, not inhibited Australia's responses to natural hazards or other civil emergencies, but it does leave gaps in the country's emergency arrangements that may, in due course, be revealed as vulnerabilities and which have been dealt with in other jurisdictions.

For example, the Civil Contingencies Act 2004 (UK) imposes an obligation upon all government departments to assess their risk and plan for their response to a foreseeable emergency and empowers the relevant minister to set standards for the performance of their duties in this regard.[14]

Canada has a Minister of Public Safety and Emergency Preparedness who is required to exercise 'leadership relating to emergency management in Canada by coordinating, among government institutions and in cooperation with the provinces and other entities, emergency management activities'.[15] The Canadian Act requires each minister to 'identify the risks that are within or related to his or her area of responsibility' and to ensure that there are in place emergency plans that are maintained and tested.[16]

In the US the Secretary of Homeland Security is 'the principal Federal official for domestic incident management'.[17] In that capacity s/he is to coordinate federal operations to prepare for, respond to, and recover from terrorist attacks, major disasters and other emergencies.[18] When the president makes an emergency or major disaster declaration, a federal coordinating officer is to be appointed to work in the disaster-affected area.[19] The federal coordinating officer is supported by Emergency Support Teams and the head of any federal agency is required to release staff to the support team, as required by the president.[20]

The president may require the governor of the affected state to appoint a state coordinating officer to work with the federal coordinating officer to ensure that the response efforts are aimed at achieving common goals.[21]

In Australia the Commonwealth Attorney-General is responsible for the Commonwealth's counter-disaster arrangements.[22] Notwithstanding this, the Attorney-General has no statutory mandate to require other government departments, responders or the private sector to prepare for emergencies nor is there a specific power to coordinate the Commonwealth response to an emergency. The Attorney-General has no specific mandate or authority to bring the states and their agencies together or to require Commonwealth agencies to release resources for the response to an emergency.

The Commonwealth response to natural disasters that overwhelm state resources but which are still managed by the affected state is managed in accordance with the Commonwealth Disaster Plan (COMDISPLAN).[23] A request for Commonwealth assistance is made by the affected state and must be approved by the Attorney-General. Once approval is obtained, the 'appropriate Australian Government agency is requested', not required, to provide the assistance and will only do so when internal approval is obtained from the minister or his or her delegate.[24] When federal assets are released to assist the states, they 'remain the responsibility of the parent agency'.[25] Each agency is to appoint Emergency Management Liaison Officers to provide a point of contact between the agency and Emergency Management Australia (EMA)[26] (an administrative unit within the Attorney General's department[27]). Emergency Management Australia may request, but not require, an Emergency Management Liaison Officer to be released to travel to the affected state 'to assist in the co-ordination of Australian Government physical assistance'.[28] EMA may also request, but not require, the affected state to provide a liaison officer to facilitate Commonwealth/state coordination.[29]

A catastrophic natural disaster will be 'an extreme hazard event that affects one or more communities, resulting in widespread, devastating, economic, health, social and environmental consequences'.[30] What distinguishes a catastrophic event from other disasters is that it will overwhelm the affected government to the extent that it is not able 'to carry out its emergency management responsibilities in the response to and/or recovery from the disaster'.[31] In such a case Commonwealth and inter-state assistance will be required not only to manage the emergency but to restore effective government. The Commonwealth's response to these catastrophic events is managed in accordance with the *National Catastrophic Natural Disaster Plan* or NATCATDISPLAN.[32] During a catastrophic disaster, the Commonwealth will 'organise to provide and/or co-ordinate assistance, upon request from the affected jurisdiction(s)'.[33] A Commonwealth coordinator 'may be appointed to support the jurisdiction's administration to ensure that response and recovery needs are met'.[34]

The Commonwealth's response to a natural hazard can be compared with the much greater role that will be assumed in responding to a terrorist incident. In the event of a terrorist incident the National Counter-Terrorism Plan provides for the declaration of a 'National Terrorist Situation'.[35] The factors that are taken into consideration in determining whether a terrorist event qualifies as such a situation are the:

- scale and nature of the incident, including the use of chemical, biological, radioactive or nuclear materials;
- involvement of multiple jurisdictions;
- involvement of Australian Government interests;
- significance of the threat;
- impact of a threat involving civil aviation or maritime operations;

- involvement of critical infrastructure; and
- involvement of foreign or international interests.[36]

Where there is a 'National Terrorist Situation', it is envisaged that the Commonwealth will take on a key leadership and coordination role. The Commonwealth will take on 'overall responsibility for policy and broad strategy'.[37] This may require it to determine 'overall policy objectives' as well as 'setting priorities between policy objectives where resources are inadequate, pre-positioning resources, international liaison, and determining public communication messages.'[38] There is nothing inherent in terrorism that brings it within the Commonwealth sphere or, when it comes to response and recovery, sets it aside from natural disasters.[39] Notwithstanding this, there is no equivalent range of powers that the Commonwealth can or will exercise in the event of a natural disaster of national consequence.[40] A similar Commonwealth response would be appropriate following a national, natural disaster.[41]

It follows that preparing for, and responding to, a natural disaster is not mainstreamed into the Australian Commonwealth government and its agencies. Emergency (other than terrorism) management remains siloed with the Attorney-General who, alone, is given responsibility for emergency planning. Commonwealth response to a disaster is dependent upon a state request and favourable responses from other ministers and departments. There is no single Commonwealth officer charged with coordinating and managing the Commonwealth's response to a disaster and the Commonwealth has no national coordination or other leadership role. When the event becomes a 'catastrophic' natural disaster, Commonwealth leadership moves from the Attorney-General to the prime minister but the COMDISPLAN remains in force, still leaving it to the attorney and other ministers to agree on what Commonwealth resources are to be released to the affected state. As the Strategic Policy Institute said:

> The Commonwealth agency seen to be responsible for [Commonwealth disaster response] . . . Emergency Management Australia (EMA), has no mandate, legislation or Cabinet endorsement with which to take command. The delivery of EMA functions for the most part is the result of goodwill on behalf of other agencies. This is clearly not a satisfactory situation.[42]

As for the executive power of the Commonwealth,[43] in the absence of emergency management legislation, the Commonwealth has no emergency power or powers. A government faced with an emergency of catastrophic proportions requires powers that would allow it to take immediate, urgent action that may not be justified in the normal course of events. Such action must be justified by law, whether that law is provided in statute or common law. Lee says:

> When a natural disaster occurs, some person needs to be put in charge of the site to direct the counter-disaster operation. It is essential that the person be conferred with extraordinary legal powers to enable him to discharge his responsibilities. For instance, he may need to authorise the entry to any land, building or structure, or cause any road or public place to be closed or to pull down, cut or remove fences, or pull down or remove buildings that may be unsafe, or burn, plough or clear firebreaks on any land, or take possession or control of any vehicles or equipment that may be necessary.[44]

The New Zealand Law Commission said:

> Emergencies are likely to call for immediate and drastic action. It follows that legislation authorising an appropriate response should be in place in advance of the emergency itself.

> This factor, and the likelihood that the emergency response will involve interference with established rights, points to the desirability of preparing emergency legislation at leisure rather than under the pressure of an actual or imminent emergency.[45]

In the absence of counter-disaster legislation there is no process for a formal declaration of a disaster or emergency at the national level, and no clear authorisation to waive the application of the 'normal' law or to take extraordinary action that is warranted by the emergency. The Commonwealth may be forced to rely on the historical prerogative power of the Crown, now encompassed in the phrase 'the Executive power of the Commonwealth' and provided for in section 61 of the Australian Constitution. The executive power of the Commonwealth:

> enables the Crown to undertake all executive action which is appropriate to the position of the Commonwealth under the Constitution and to the spheres of responsibilities vested in it by the Constitution. It includes the prerogative powers of the Crown, that is the powers accorded to the Crown by common law.[46]

The scope of the prerogative power is uncertain,[47] and resists being defined as a list of powers or subject areas.[48] The prerogative power has included a power vested in the Crown to respond to emergencies that are 'a national emergency, [where there is] an urgent necessity for taking extreme steps for the protection of the Realm'.[49] Lee, in his review of emergency powers, said:

> a special or emergency prerogative lies dormant in the fabric of executive powers. Such a prerogative awaits activation in the face of extreme necessity. The submission in this work is that the Commonwealth possesses a prerogative power to requisition a subjects' property... Another assertion... is that a case can be made for an extraordinary prerogative which extends to the assumption of legislative power when the legislative arm of government is paralysed.[50]

Renfree states:

> The prerogative of the Crown... arises from a general principle that in time of emergency the law arms Crown and subject alike with the right of intervening, and sets public safety above private right... Apart from natural disasters and political crises, there are two main crises that may confront a nation – attack from abroad and domestic violence within.[51]

Viscount Radcliffe thought 'Riot, pestilence and conflagration might well be other circumstances' that would trigger the use of the prerogative.[52]

In the UK the Civil Contingencies Act 2004 specifically refers to the Royal Prerogative when identifying the permissible subject areas for emergency regulations.[53] This provision reflects a view that the prerogative continues to apply and can provide a lawful basis to exercise necessary emergency powers to deal with civil contingencies such as natural disasters.

The executive power of the Commonwealth, which includes the ancient prerogative powers, has had to be understood in the context of Australia's constitutional, federal arrangements. Part of the power, in Australia, includes an implied 'nationhood' power – the power to the Commonwealth government to act, or react, to situations which, because of their size or nature, require a single national response. In *Davis v. Commonwealth*, Brennan CJ said:

> s.61 [of the Constitution] does confer on the Executive Government power 'to engage in enterprises and activities peculiarly adapted to the government of a nation and which cannot

> otherwise be carried on for the benefit of the nation' ... *It invites consideration of the sufficiency of the powers of the States to engage effectively in the enterprise or activity in question and of the need for national action (whether unilateral or in co-operation with the States) to secure the contemplated benefit.*[54]

In 2009 Australia, along with the rest of the world, was facing the global financial crisis. In response the Commonwealth government proposed a stimulus package that involved payments to individuals, based on their taxable income, ranging from $250 to $900. The payments were to be used to encourage consumption and to maintain economic activity. The payments were to be made from Consolidated Revenue. A taxpayer sought a declaration that the payments were unlawful. The Commonwealth argued that the Constitution allowed the government to appropriate money from Consolidated Revenue for 'the purposes of the Commonwealth'.[55] The plaintiff argued that, absent a specific power, issuing the stimulus payment was not a 'purpose of the Commonwealth'. Further, the passage of an Act to appropriate money for a particular purpose could not make that purpose a 'purpose of the Commonwealth'. What was a legitimate purpose of the Commonwealth had to be determined by reference to the federal structure enshrined in the Constitution and by the specific list of subject areas within the legislative power of the Commonwealth.[56]

French CJ took the view that the economic stimulus package was a legitimate exercise of the nationhood power. He said:

> The executive power extends, in my opinion, to short-term fiscal measures ... where such measures are on their face peculiarly within the capacity and resources of the Commonwealth Government ... In this case the Commonwealth had the resources and the capacity to implement within a short time-frame measures which, on the undisputed facts, were rationally adjudged as adapted to avoiding or mitigating the adverse effects of global financial circumstances affecting Australia as a whole, along with other countries.[57]

Gummow, Crennan and Bell JJ held that the financial crisis was analogous to a natural disaster:

> Determining that there is the need for an immediate fiscal stimulus to the national economy in the circumstances set out above is somewhat analogous to determining a state of emergency in circumstances of a natural disaster. The Executive Government is the arm of government capable of and empowered to respond to a crisis be it war, natural disaster or a financial crisis on the scale here. This power has its roots in the executive power exercised in the United Kingdom up to the time of the adoption of the Constitution but in form today in Australia it is a power to act on behalf of the federal polity.[58]

The decision is not conclusive;[59] but four of the seven judges upheld the concept that the executive power of the Commonwealth would extend to allow it to take extraordinary measures in response to a truly national emergency including, arguably, an overwhelming natural disaster.

History shows that the Commonwealth has exercised such power. In 1974 the Director-General of the Commonwealth's Natural Disasters Organisation (now EMA) was appointed to take supreme command of recovery operations following the devastation of Darwin by Cyclone Tracey. The appointment of the Commonwealth officer as supreme commander was 'because the situation in Darwin was a national disaster of major dimensions'.[60] Given that Darwin was in the Northern Territory, the government could have relied on its power to make laws with respect to territories as, at that time, the Northern Territory was not a self-governing territory but was administered by the Commonwealth government.[61] That was not, however, (according to Major General Stretton[62]) the basis of the appointment. If the Commonwealth did not purport to rely on the territories power and, in the absence of any other constitutional head of power, or should a similar situation arise in a state, then any power of the Commonwealth to manage 'a national disaster of major dimensions' must be an exercise of the executive or prerogative power of the Commonwealth.[63]

On the other hand, the fact that three judges in *Pape's case* did not see the executive power as providing the constitutional basis to the global crisis shows that the true nature and extent of the Commonwealth executive power is not clear or without controversy.[64] Where a government purports to rely on the prerogative power, there may be challenges as to whether the power exists,[65] and whether it extends to allow the measures that the government proposes.[66] Leaving the matter up to a court to determine when and how the Commonwealth may act in an emergency would not be appropriate when such powers are required as a matter of urgency.

The New Zealand Law Commission recommended against relying on non-statutory authority to manage disaster response, because it is 'vague and ill defined' and is not subject to scrutiny.[67] The Commission argued that it is better to plan for the possible eventualities and to define in statute, before a disaster strikes, what the scope of necessary emergency powers will be.[68]

Lee argues that 'emergency provisions should be embodied in legislation which makes their existence conspicuous'.[69] He notes that, following the bombing of the Hilton Hotel in 1978, the then Leader of the Opposition (and later Governor-General), Mr Hayden, argued for Commonwealth emergency legislation:

> not so much in order to confer sweeping new powers but rather to circumscribe, confine and define their exercise, and to remove some of the extraordinary uncertainties which now prevail.[70]

It appears that the Commonwealth should enact legislation, before disaster strikes, to ensure that the necessary emergency powers are in place, with clear criteria for when they may be used.[71]

The states

The Australian states and territories have well developed civil contingencies legislation that establishes the necessary response agencies and provides for emergency powers and, if necessary, for the mobilisation of the whole of government to respond to any type of emergency.

Each state has its own police force,[72] a State Emergency Service for managing the response to floods, storms and tsunami,[73] and fire fighting services.[74] Each state also has an emergency plan and, except in the Australian Capital Territory, there are arrangements for regional and local plans. Specialised councils and committees exist to advise the relevant minister on specific hazards, in particular flood and bushfire management. In most jurisdictions there are plans to bring together the whole of cabinet to coordinate the response across the whole of government.[75]

The disaster plan for each jurisdiction nominates a particular agency to be in control of a particular incident. These combat, management, lead, controlling, control or hazard management agencies have the principal responsibility to respond to the nominated hazard event and to control the response of the other responders.[76] As the situation escalates, management of the response may move to regional and ultimately state control and coordination. There is a standing appointment of local, regional and state emergency controllers (usually police officers) who can take control of the response arrangements should the situation require a more coordinated response than can be provided by first responders.[77] There is also a trend to establish standing recovery controllers who can take over and manage the long-term recovery from the impact of a natural disaster.[78] These arrangements are similar to those in the UK, where Regional Controllers can be appointed,[79] and where lead agencies are nominated to take control of the response to particular hazard events.[80]

The Civil Contingencies Act 2004 (UK) provides that emergency regulations can be made when, inter alia, 'an emergency has occurred, is occurring or is about to occur'.[81] These regulations may give a minister or a Regional Emergency Coordinator extensive power to restrict movement and private rights in the community's interest of responding to the disaster.[82] These powers are in addition to those powers given to the first responders as part of their normal duties.[83] In Australia extensive powers are vested in the relevant minister and/or the state coordinator and are largely spelt out as standing powers – that is, they are already set out in the relevant legislation. This allows agencies to plan for, and understand in advance, what powers they will have during a declared state of emergency, disaster or alert.[84] There is in each case a general power to do whatever is necessary or make such orders as are required to ensure an effective response, but the general powers to require people to evacuate, to commandeer private assets, to require people to assist in the emergency response or to take and use water and the like are set out well in advance of any particular emergency.

Local government

Unlike in the UK local authorities are given a very limited express role in emergency management and emergency response. In the UK local authorities are category 1 responders, with obligations to plan for and respond to emergencies within their area.[85] In a draft report, *Barriers to Effective Climate Change Adaptation*, the Australian Productivity Commission said:

> local governments are primarily responsible for managing disasters in their local government area, while other agencies such as the police and State Emergency Service perform emergency management activities on a larger scale under command structures determined at the state or territory level.[86]

That does not, and did not, reflect the law in most Australian states. For example, in New South Wales the specific obligations upon local authorities are to pay contributions toward the costs of maintaining the rural and urban fire brigades and the State Emergency Service;[87] to recommend a person for appointment as the local State Emergency Service controller;[88] and generally to 'prepare for emergencies'.[89]

In preparing for emergencies a New South Wales local government authority is to provide executive support for the Local Emergency Management Committee (LEMC) and the general manager of the local government authority is to chair the LEMC. The general manager is also a member of the Regional Emergency Management Committee, which is chaired by a police officer appointed as the Regional Emergency Operations Controller.[90]

A local authority *may* form a rural fire brigade if there is a rural fire district within their area.[91] The authority is to provide necessary facilities and accommodation to allow a Fire Control Officer, an employee of the Rural Fire Service, not the council, to perform his or her duties.[92] The authority is to appoint one nominee to sit on the local Bushfire Management Committee.[93]

A local authority is required to take any steps to prevent bushfires starting on, or spreading from its land.[94] Where a local authority is aware of a fire burning on its land during a bush fire danger period, they must take steps to extinguish the fire and, if they cannot extinguish it, notify a rural fire brigade, fire control officer or other prescribed officer.[95] A council is required to conduct building inspections to determine if adequate provision for fire safety has been made,[96] and it must monitor compliance with fire safety standards.[97] Councils must develop a coastal zone management plan that includes provision for 'emergency actions carried out during periods of beach erosion'.[98]

It is a function of local governments in South Australia 'to take measures to protect its area from natural and other hazards and to mitigate the effects of such hazards',[99] which imposes an obligation to plan for hazard response.

Local government in Tasmania provide the chair of each Municipal Emergency Management Committee and must nominate a person as Municipal Coordinator. They must maintain and provide resources for emergency response in accordance with the Municipal Emergency Management Plan and establish and maintain sufficient units of the State Emergency Service (SES). They must, in agreement with the Director of the SES, provide accommodation and facilities to allow the SES unit to undertake training and store equipment and resources.[100]

Victorian councils are required to appoint a Municipal Emergency Planning Committee, which is required to prepare a draft plan for consideration and adoption by the council.[101] The local plan must identify the municipal resources available for emergency response and recovery and list any 'neighbourhood safer places' and/or community fire refuges that have been identified within the local government area.

A Western Australian local government authority is required to ensure that effective local emergency management arrangements are in place and is to manage the recovery process following an emergency.[102]

There is no specific role provided for local government in emergency management in the Northern Territory, although each local government authority is to nominate a person to sit on the Regional Emergency and Regional Recovery committees.[103]

Queensland is the notable exception, providing a central role for local government in emergency management. It is an express principle that, in Queensland, 'local governments should primarily be responsible for managing events in their local government area'.[104] It is local government that has the majority representation in planning for disasters and it is local government, rather than the minister or a state committee, which must approve the local disaster management plan. A local government authority must ensure that it has the ability to provide equipment and people to deal with an emergency situation or disaster, or to help another agency to deal with the situation.[105]

Reflecting this reality, the Productivity Commission changed the text in its final report from 'local governments are primarily responsible for *managing* disasters'[106] to the more accurate 'local governments are primarily responsible for *supporting* disaster management'.[107]

What's missing?

What is missing from Australian emergency, or civil contingencies, management is a whole-of-government and whole-of-community approach. As discussed above, the primary response and management roles are located at state level. The Commonwealth has only limited roles in responding to an emergency, including a national emergency, and its powers are undefined. Local government remains subordinate to state government and is given little authority to take the lead in developing local community resilience. Where there are legal arrangements for emergency or civil contingency management these identify how and upon whose authority government will act – they describe and authorise the government's procedures rather than a whole-of-community response.

As discussed, Australia's federal structures require cooperative governance. Neither the states nor the Commonwealth are subordinate to the other; they must work together to achieve community objectives. The latest development in the area of cooperative governance is the *National Strategy for Disaster Resilience*, a strategy adopted by COAG and representing the consensus of all levels of government.[108] The *National Strategy* recognises that emergency management is not simply a government function but that it requires contributions from the private and not-for-profit sector and from communities and individuals. The *National Strategy* commits all levels of government to work together to develop community resilience to natural disasters and to share responsibility for all aspects of emergency management. The brief review of Australia's emergency management arrangements set out here would suggest that there is still some way to go.

Some issues that need to be addressed to advance the *National Strategy* are:

- defining what resilience means in the context of the variety of Australian communities;
- how resilience can be measured or assessed to know whether the objectives of the strategy are being met and what would constitute a successful implementation;
- developing contingency response plans in cooperation with non-government sectors in order to include all aspects of the community into emergency planning and response arrangements; emergency plans have to move from being 'government' plans to whole-of-community plans.

Conclusion

This paper has reviewed the arrangements to manage an emergency or 'civil contingency' in Australia. This review contributes to the discussion in this volume identifying the structural arrangements for emergency management and response. Australia's emergency management arrangements have generally served the community well but there are undoubted limitations. One identified here is the lack of established Commonwealth powers. This has not proved problematic to date, but preparation for a catastrophic event, of national significance, should not be avoided simply because it has not yet happened. Threats posed by climate change, increasing populations and the desire of Australians to live along the thin coastal strip will expose the population to more and more extreme hazard events.

The other limitation identified has been the nature of current arrangements, which focus on governments and their response to emergencies. They have, to date, largely omitted other players, in particular non-government actors, whether that is private industry, communities or individuals. There is, however, evidence that this is changing with the *National Strategy for Disaster Resilience*. All the same, the *National Strategy* is an overarching policy statement. What is required now is work at the local level to develop true local resilience to the impact of inevitable natural hazards such as fire, flood and storm.

Notes

1. See, in particular, Australian Constitution s 51.
2. *Commonwealth v. Tasmania* (1983) 158 CLR 1 ('Tasmanian Dam case').
3. *New South Wales v. Commonwealth* (2006) 229 CLR 1.
4. Australian Constitution s 109.
5. Ibid., s 122.
6. Australian Capital Territory (Self-Government Act) 1988 (Cth); and Northern Territory (Self-Government Act) 1978 (Cth).
7. *The Commonwealth v. Australian Capital Territory* [2013] HCA 55 (12 December 2013).
8. Local Government Act 1993 (NSW); Local Government Act (NT); Local Government Act 2009 (Qld); Local Government Act 1999 (SA); Local Government Act 1993 (Tas); Local Government Act 1989 (Vic); and Local Government Act 1995 (WA).
9. Department of Infrastructure and Regional Development, *Local Government*, 27 November 2013, http://www.regional.gov.au/local/.
10. Council of Australian Governments, 'Welcome to the Council of Australian Governments (COAG) website', http://www.coag.gov.au/ (accessed December 14, 2013).
11. Australian Constitution s 51.
12. Attorney-General's Department, *National Emergency Management Projects*, http://www.em.gov.au/Fundinginitiatives/NationalEmergencyManagementProjects/Pages/default.aspx (accessed December 14, 2013); Attorney-General's Department, *Natural Disaster Relief and Recovery Arrangements*, http://www.em.gov.au/Fundinginitiatives/Naturaldisasterreliefandrecoveryarrangements/Pages/default.aspx (accessed December 14, 2013); and Social Security Act 1991 (Cth), Part 2.23B Disaster Recovery Allowance, 2.24 Australian Government Disaster Recovery Payment.
13. Attorney-General's Department, *Australian Government Disaster Response Plan* (COMDISPLAN) (Canberra: Commonwealth of Australia, 2008).
14. Civil Contingencies Act 2004 (UK) s 2.
15. Emergency Management Act, SC 2007, c 15, s 3.
16. Ibid., s 6.
17. Homeland Security Presidential Directive/HSPD-5, 28 February 2003, para 4.
18. Ibid.
19. Robert T. Stafford Disaster Relief and Emergency Assistance Act 42 USC 5143.
20. Ibid., 5144.
21. Ibid., 5143.
22. Commonwealth of Australia, *Administrative Arrangements Order*, 18 September 2013.
23. COMDISPLAN.
24. Ibid., para 4.2.3.
25. Ibid., para 4.6.1.
26. Ibid., para 4.8.1.
27. Attorney-General's Department, *Organisational Structure*, http://www.ag.gov.au/About/Documents/Attorney-Generals%20Department%20Organisational%20Chart.PDF (accessed December 14, 2013).
28. COMDISPLAN, para 4.8.
29. Ibid.
30. Attorney-General's Department, *National Catastrophic Natural Disaster Plan* (NATCATDISPLAN) (Canberra: Commonwealth of Australia, 2010).
31. Ibid., para 6.
32. Ibid., para 2.
33. Ibid., para 7.
34. Ibid., para 19.
35. *National Counter-Terrorism Plan*, 2nd ed. (Canberra: Commonwealth of Australia, 2005), para 85.
36. Ibid., para 86.
37. Ibid.
38. Ibid.
39. David Templeman and Anthony Bergin, *Taking a Punch: Building a More Resilient Australia* (Canberra: Australian Strategic Policy Institute, 2008), 8.

40. Attorney General's Department, *Australian Emergency Management Arrangements* (Canberra: Commonwealth of Australia, 2009), 19.
41. Anthony Bergin and David Templeman, 'Get ready for the Big One', *The Australian* (Sydney), March 10, 2009, 22.
42. Templeman and Bergin, *Taking a Punch*, 7.
43. The discussion in this section is based on work for Michael Eburn, 'Australia's International Disaster Response: Laws, Rules and Principles' (PhD diss., Monash University, 2009), published as Michael Eburn, *Australia's International Disaster Response – Laws, Rules and Principles* (Saarbrucken: VDM-Verlag, 2010). See also Eburn, 'Responding to Catastrophic Natural Disasters and the Need for Commonwealth Legislation', *Canberra Law Review* 10, no. 3 (2011): 81.
44. H.P. Lee, *Emergency Powers* (Sydney: Law Book Company, 1984), 171–2.
45. New Zealand Law Commission, *Final Report on Emergencies* (Wellington: Government of New Zealand, 1991), para 4.12.
46. *Barton v. The Commonwealth* (1974) 131 CLR 477, 498 (Mason J).
47. *Burmah Oil Co. Ltd. v. Lord Advocate* [1965] AC 75: 99 (Lord Reid), 145 (Lord Pearce); H.E. Renfree, *The Executive Power of the Commonwealth of Australia* (Sydney: Legal Books, 1984), 389, 394; and New Zealand Law Commission, *Final Report on Emergencies*, paras 4.37–4.41.
48. Tony Blackshield and George Williams, *Australian Constitutional Law and Theory*, 4th ed. (Sydney: Federation Press, 2006), 523, 525; and *Burmah Oil Co. Ltd. V. Lord Advocate* [1965] AC 75: 114 (Viscount Radcliffe).
49. *Burmah Oil Co. Ltd. V. Lord Advocate* [1965] AC 75: 136 (Lord Hodson).
50. Lee, *Emergency Powers*, 322.
51. Renfree, *The Executive Power of the Commonwealth of Australia*, 466. See also George Winterton, 'The Limits and Use of Executive Power by Government', *Federal Law Review* 31 (2003): 421, 425.
52. *Burmah Oil Co. Ltd. V. Lord Advocate* [1965] AC 75: 115 (Viscount Radcliffe).
53. Civil Contingencies Act 2004 (UK) s 22(3).
54. *Davis v. Commonwealth* (1988) 166 CLR 79: 111 (Brennan J) (emphasis added).
55. Australian Constitution s 81.
56. Ibid., s 51.
57. *Pape v. Commissioner of Taxation* (2009) 238 CLR 1: para 133 (French CJ).
58. Ibid., para 233 (Gummow, Crennan and Bell JJ).
59. Hayne and Keiffel JJ would have upheld the law, at least in part, as a law with respect to taxation; Heydon J dissented, finding the law was not a valid law of the Commonwealth.
60. Alan Stretton, *Darwin Disaster – Cyclone Tracy: Report by Director-General Natural Disasters Organisation on the Darwin Relief Operations 25 December 1974–3 January 1975* (Canberra: Commonwealth of Australia, 1975), para 8.
61. Australian Constitution s 122.
62. Stretton, *Darwin Disaster*. See also Stretton, *The Furious Days: The Relief of Darwin* (Sydney: Collins, 1976); Stretton, *Soldier in a Storm: An Autobiography* (Sydney: Collins, 1978); and Lee, *Emergency Powers*, 322.
63. Australian Constitution s 61.
64. *Pape v. Commissioner of Taxation* (2009) 238 CLR 1.
65. *Attorney General v. De Keyser's Royal Hotel* [1920] AC 508; and *Ruddock v. Vadarlis* (2001) 110 FCR 491.
66. As in *Pape v. Commissioner of Taxation* (2009) 238 CLR 1.
67. New Zealand Law Commission, *Final Report on Emergencies*, para 1.29.
68. Ibid.
69. Lee, *Emergency Powers*, 193.
70. Ibid., 192.
71. Eburn 'Responding to Catastrophic Natural Disasters', 81.
72. Police Act 1990 (NSW); Police Administration Act (NT); Police Service Administration Act 1990 (Qld); Police Act 1998 (SA); Police Service Act 2003 (Tas); Police Regulation Act 1958 (Vic); and Police Act 1892 (WA). Policing services in the Australian Capital Territory are provided by the Australian Federal Police established by the Australian Federal Police Act 1979 (Cth).
73. Emergencies Act 2004 (ACT); State Emergency Service Act 1989 (NSW); Disasters Act 1982 (NT); Disaster Management Act 2003 (Qld); Fire and Emergency Services Act 2005 (SA);

Emergency Services Act 1976 (Tas); Victoria State Emergency Service Act 2005 (Vic); and Fire And Emergency Services Act 1998 (WA).

74. Emergencies Act 2004 (ACT); Fire Brigades Act 1989 (NSW); Rural Fires Act 1997 (NSW); Fire and Emergency Act 1996 (NT); Bushfires Act 1980 (NT); Fire and Rescue Service Act 1990 (Qld); Fire and Emergency Services Act 2005 (SA); Fire Service Act 1979 (Tas); Metropolitan Fire Brigades Act 1958 (Vic); Country Fire Authority Act 1958 (Vic); Bush Fires Act 1954 (WA); and Fire Brigades Act 1942 (WA).
75. Michael Eburn, *Emergency Law*, 4th ed. (Sydney: Federation Press, 2013), chap. 5.
76. State Emergency and Rescue Management Act 1989 (NSW) s 3; Department of Police and Emergency Management, *Tasmanian Emergency Management Plan (Issue 7)* (Hobart: Government of Tasmania, 2012); State Disaster Management, *Group Interim Queensland State Disaster Management Plan 2012* (Brisbane: Government of Queensland, 2012); Northern Territory Emergency Service, *Northern Territory All Hazards Emergency Management Arrangements* (Darwin, 2011); State Emergency Management Committee *State Emergency Management Plan – Public Version* (Adelaide: Government of South Australia, 2012); and Emergency Management Act 2005 (WA) s 4.
77. Emergencies Act 2004 (ACT); State Emergency and Rescue Management Act 1989 (NSW); Disasters Act (NT); Disaster Management Act 2003 (Qld); Emergency Management Act 2004 (SA); Emergency Management Act 2006 (Tas); Emergency Management Act 1986 (Vic); and Emergency Management Act 2005 (WA).
78. See, for example, State Emergency and Rescue Management Act 1989 (NSW) ss 20A, 20B; Emergency Management Bill 2013 (NT); Disaster Management Act 2013 (Qld) s 21D; and Emergency Management Act 1986 (Vic) ss 17A-17F.
79. Civil Contingencies Act 2004 (UK) s 24.
80. The Civil Contingencies Act 2004 (Contingency Planning) Regulations 2005 (UK) SI 2005/2042 paras 9–10.
81. Civil Contingencies Act 2004 (UK) s 21(2).
82. Ibid., s 22(3).
83. See, for example, Fire and Rescue Services Act 2004 (UK) s 44.
84. See fn 77.
85. Civil Contingencies Act 2004 (UK) s 2, Schedule 1.
86. Productivity Commission, *Draft Report: Barriers to Effective Climate Change Adaptation* (Canberra: Commonwealth of Australia, 2012), 197–8.
87. Fire Brigades Act 1989 (NSW) ss 45–56; Rural Fires Act 1997 (NSW) ss 101–114; and State Emergency Service Act 1989 (NSW) ss 24A–24M.
88. State Emergency Service Act 1989 (NSW) s 17.
89. Local Government Act 1993 (NSW) s 22.
90. State Emergency and Rescue Management Act 1989 (NSW) ss 22–32.
91. Rural Fires Act 1997 (NSW) s 15.
92. Ibid., s 37.
93. Ibid., s 50; and Rural Fires Regulation 2008 (NSW) cl 14.
94. Rural Fires Act 1997 (NSW) s 63.
95. Ibid., s 64.
96. Environmental Planning and Assessment Act 1979 (NSW) s 118M.
97. Environmental Planning and Assessment Regulation 2000 (NSW) Part 9.
98. Coastal Protection Act 1979 (NSW) ss 55C-55D.
99. Local Government Act 1999 (SA) s 7(d).
100. Emergency Management Act 2006 (Tas) ss 46–49.
101. Emergency Management Act 1986 (Vic) ss 20, 21.
102. Ibid., s 41.
103. Emergency Management Act 2013 (NT) ss 61, 72.
104. Disaster Management Act 2003 (Qld) s 4A.
105. Ibid., s 80.
106. Productivity Commission, *Draft Report*, 197–8 (emphasis added).
107. Productivity Commission, *Barriers to Effective Climate Change Adaptation* (Canberra: Commonwealth of Australia, 2013), 247 (emphasis added).
108. Council of Australian Governments, *National Strategy on Disaster Resilience* (Canberra: Commonwealth of Australia, 2012).

The power to react: review and discussion of Canada's emergency measures legislation

John Lindsay

Department of Applied Disaster and Emergency Studies, Brandon University, Brandon, Canada

Almost a century ago Canada implemented the War Measures Act to give the federal government extraordinary powers to deal with the domestic implications of the First World War. The government further expanded and institutionalised these powers through the Second World War, the Korean crisis and the cold war period. Now these war emergency powers are entrenched in both federal and provincial emergency measures legislation and exercised for peacetime disasters. This paper reviews the historical progression of these powers to better understand the original context for their development and to connect past abuses of civil rights under the War Measures Act to the protections and limits incorporated in the current legislation. Then, based on this understanding of where the powers originated, the paper takes a comprehensive look at the powers available across the Canadian jurisdictions. It identifies common groups of powers and discusses regional variations. It concludes with a short discussion on the imbalance between extraordinary powers for response and the lack of a corresponding emphasis on reducing the risks that may generate the need for them.

Introduction

Extreme, damaging events, of both natural and technological origins, occur in Canada every year. However, not all are declared to be disaster events under the relevant provincial and federal legislation. In this sense disasters in Canada are rare events. This is an important factor when considering how Canada's current emergency management legislation is written and why its main focus remains on preparing for response while undervaluing mitigation and recovery.

There has never been a declared National Emergency under the Emergencies Act 1988.[1] There are, by contrast, several State of Emergency declarations every year across the country under the authority of the various provincial acts. While the Canadian Disaster Database,[2] maintained by Public Safety Canada, does not differentiate events that generated a legal declaration from other events, it is safe to assume that many of the 224 natural and technological events registered between 2000 and 2010 would have required, or at least qualified for, a local State of Emergency.

Canada is a geographically large nation with a relatively small population concentrated in urban centres and along the southern portion of the country. This limits the potential for large-scale emergencies, as much of the Canadian landscape is too sparsely populated and underdeveloped to generate significant losses of life or property in the event of a natural or technological hazard impact. Of the 224 events between 2000 and 2010 only eight caused 10 or more fatalities and only one event, the 14 July 2000 tornado in Pine Lake Alberta, generated those fatalities in an acute hazard impact, the others being either disease outbreaks or collections of deaths during a particular season, such as the 2003 avalanche season in Western Canada.

One implication of this lack of experience is that disasters do not create a significant political issue in Canada. The way disasters are managed, both in practice and in the legislation that empowers and restricts that practice, is not an immediate concern of the resident or the elected official – or perhaps more importantly of the voter or the candidate. Canada's federal emergency management legislation is fairly recent, being first passed in 1988. Many provinces can trace their legislation back to the 1950s, but complete revisions are rare. Managing hazards, preparing and responding to impacts and providing for the long-term recovery of communities could not be judged to be a priority of Canadian governments based on the history of the legislation.

Furthermore, the legislative approach to emergency management, especially the powers available under a state of emergency, are rooted in Canada's wartime experiences. The legislation, and the government agencies that administer that legislation, come to the issue largely from a security perspective. Recovery and mitigation are often overlooked as a result of this reactive, response-oriented and state-centric approach. While some of the legislation has civil liberty protection incorporated, that perspective is not universally considered, and it certainly does not go as far as Geis's desire for a human right to safety.[3]

This paper will first discuss Canada's Constitution and five significant historical influences to establish the foundation for an examination of the federal and provincial legislation of relevance to emergencies. Then a review of the legislation will consider the comprehensiveness of the powers and the protection provided for civil rights. This will inform a series of conclusions about Canada's emergency management law and pose challenges for its future development.

Historical roots of emergency powers

Canada's emergency management law has been shaped by two main forces: the division of powers laid out in the British North America Act of 1867, which was later repatriated to be Canada's Constitution of 1982,[4] and the experience of emergency powers during past crises. The latter can be further divided into five influential experiences: the domestic events of the First and Second World Wars, the tensions of the cold war period, the unique circumstances of the Front de libération du Québec (FLQ) crisis and the more recent fallout from Y2 K (i.e. the turn of the millennium), the 11 September 2001 terrorist attacks and severe acute respiratory syndrome (SARS). These five have all contributed to the Canada's legislative framework through the common filter of the Constitution.

Canada's Constitution

There can be no meaningful discussion of Canada's emergency management legislation without a basic appreciation of the impact of Canada's Constitution. 'The validity of federal emergency legislation has always turned on the question whether the legislation was a valid exercise of the federal emergency power or whether it unnecessarily invaded a field reserved to the provinces under s.92 of the [British North America] Act.'[5] While this question of legal validity has been played out in the Supreme Court, usually in the wider context of the often tumultuous federal–provincial relationship, in practice emergency management is handled by the Provinces through the application of their own legislation. However, Canada does have legislation at both levels; these laws are designed, with varying effectiveness, to interact.

Canada's first Constitution, drafted as the British North America Act 1867 by the UK Parliament, reflects the issues of the time, including the Civil War in the US. One outcome of this setting was that 'the provinces were to be allotted exclusive legislative powers over enumerated classes of matters and to the federation should belong the vast residue of undefined legislative power'.[6] These powers dealt with matters that were common at the time and, as emergency management was not then the professional practice it is today, it was not listed as a distinct legislative subject assigned to either the federation or the provinces. However, there is a general residual power, first created in section 91 of Canada's 1867 Constitution,[7] 'to make Laws for the Peace, Order, and good Government of Canada, in relation to all Matters not coming within the Classes of Subjects by this Act assigned exclusively to the Legislatures of the Provinces' (known as the POGG powers) that has been the federal government's 'foot in the door' to emergency management in past crises and continues to underlie its involvement today.

First World War

The First World War is often seen as Canada's first real test as a nation. Battles such as Passchendaele and Vimy Ridge saw Canadian soldiers distinguish themselves and helped carve out the country's international reputation. The impact on the domestic front was no less significant as a country comprised of immigrants from around the world came to terms with what it meant to be Canadian rather than a foreigner.

In August 1914 the Government of Canada passed the War Measures Act to allow it to exert control by decree.[8] It granted the Governor in Council the following carte blanche powers in section 6 to administer the country during the crisis:

> The Governor in Council shall have power to do and authorize such acts and things, and to make from time to time such orders and regulations, as he may by reason of the existence of real or apprehended war, invasion or insurrection deem necessary or advisable for the security, defence, peace, order and welfare of Canada; and for greater certainty, but not so as to restrict the generality of the foregoing terms, it is hereby declared that the powers of the Governor in Council shall extend to all matters coming within the classes of subjects hereinafter enumerated, that is to say:
>
> (a) censorship and the control and suppression of publications, writings, maps, plans, photographs, communications and means of communication;
> (b) arrest, detention, exclusion and deportation;
> (c) control of the harbours, ports and territorial waters of Canada and the movements of vessels;
> (d) transportation by land, air, or water and the control of the transport of persons and things;
> (e) trading, exportation, importation, production and manufacture;
> (f) appropriation, control, forfeiture and disposition of property and of the use thereof.

Issuing orders and regulations became the norm for the Canadian government during this wartime period. Starting almost immediately, Orders in Council were made regarding issues such as the Registration of Alien Enemies,[9] or Trading with the Enemy.[10] The intention was to limit opportunities for espionage or for otherwise aiding the enemies of Canada. The Orders in Council had the effect of turning some residents of Canada into prisoners of war on the basis of their original nationality. The language of the day seems especially painful when read now:

> Where any alien of enemy nationality interned under the provisions of this order has wife or children living with and dependent on him, such wife and children *shall be permitted to accompany* him.[11]

As a result of this Order in Council, over 120,000 people living in Canada, mostly immigrants who had come to settle the prairies, were deemed Alien Enemies and 7762 of them served time in internment camps, often providing labour for government construction projects.[12] The orders under the War Measures Act ended in 1920.

During the 1930s there were several cases that went to the Judicial Committee of the Privy Council regarding the Federal government's use of the POGG residual power to interfere in other matters, mostly relating to employment, that were assigned to the Provinces. In their rulings their lordships defined the need for these POGG powers to be used for emergency situations when temporary measures are called for.[13] The use of these extraordinary powers was to be tested again as the world moved once more to war.

Second World War

Canada's actions in the Second World War were no less disturbing. The War Measures Act was renewed on 3 September 1939 and a newly drafted set of Regulations for the Defence of Canada was adopted by Order in Council. These 64 Regulations had been prepared in advance and covered a wide range of issues.[14] Clearly the lessons of the First World War and the advice of other nations (especially the UK, which followed the same course) had been heeded. The Regulations were designed to ensure the safety of the country during the war and took full advantage of the broad scope of power to rule by decree granted by the War Measures Act.

Many of the Regulations were reactions to a change in the nature of warfare. The Spanish Civil War had highlighted the importance of aircraft as an offensive weapon, and the prospect of aerial bombardment of not just the battlefield but of civilian targets far behind the frontlines led to the need for civil defence. The Regulations included air raid precautions and tasked provinces with administering aspects of the resulting programmes. This led to some court challenges and a series of policy adjustments to account for the costs and clarify the authority under which the Regulations would be enforced.[15] Thus, from the very beginning, the federal–provincial relationship with respect to civil defence – what is now emergency management – has been fraught with jurisdictional uncertainties.

What was not uncertain during the war was the intention of the federal government to control dissent and protect the nation from extreme ideologies from the left and the right. Fascist and Communist organisations and sympathisers were repressed as the Regulations were implemented to ensure a home front devoid of subversion or even criticism of the government. The War Measures Act provided the power to do so, and the Regulations put those powers to work. Among the worst abuses of these powers was the internment of Japanese Canadians in British Columbia. After the attack on Pearl Harbor the Canadian government, under the power of the War Measure Act, ordered the seizure of some 1800 fishing boats from Japanese Canadians and forced the evacuation of about 22,000 Japanese to labour camps. The government then seized and sold the property of those unfortunates to pay for their internment.

The Regulations for the Defence of Canada were also used to target individuals and organisations that were critical of the government. The irony, which has been recognised as recurring during the current 'War on Terror', is that it appears necessary to restrict freedoms in order to protect those very freedoms. The War Measures Act went further, however, to the point that freedoms were sacrificed at a cost that may have exceeded the benefits. Lester Phillips has reviewed the Regulations and concluded: 'In short, one may raise some doubt as to whether the Liberal war-time government has lived up to a democratic standard of keeping war-time restrictions to a minimum consonant with military necessity and proportionate to the existing danger to the nation'.[16]

Cold war civil defence and provincial legislation

The end of the Second World War may have seen the end of wartime measures in Canada but not the end of wartime thinking. Many of the activities initiated under the Regulations for the Defence of Canada were transformed from temporary reactions to institutionalised practices. The National Emergency Transitional Powers Act, 1945 was passed for one year to bridge the gap between the end of the War Measures Act and the completion of demobilisation. It also extended the powers relating to the deportation of Canadians of Japanese ancestry. These actions were challenged by the Co-operative Committee on Japanese Canadians all the way to the Judicial Committee of the Privy Council, which upheld the Orders in Council. The Act was later amended and extended by the Continuation of Transitional Powers Act 1947 so that these powers remained in effect until 1951.

The outbreak of hostilities on the Korean peninsula, and particularly the involvement of China in that conflict, led the Canadian government under Prime Minister St-Laurent to bring forward the Emergency Powers Act 1951. When introducing the bill, the prime minister explained that using the existing War Measures Act to deal with the necessary defence production and cost control issues could further the idea that war was inevitable and place unnecessary limits on parliamentary involvement. Instead the new legislation was intended to provide 'adequate powers to deal with the type of situation that may be expected to arise, the government ha[ving] endeavoured to avoid taking powers of the grave character which no democratic government wishes to have, as a government, unless those powers are really necessary for the safety of the state.'[17] The Emergency Powers Act was originally limited to one year but was continued with Parliament's consent until May 1954.

While the Korean 'international emergency' had generated an alternative approach, the War Measures Act remained part of Canada's Statutes throughout the cold war decades. It was amended in 1960 so that 'Any act or thing done or authorized or any order or regulation made under the authority of the Act, shall be deemed not to be in abrogation, abridgement, or infringement of any right or freedom recognized by the Canadian Bill of Rights'. It remained available to the government in the event a 'real or apprehended war, invasion or insurrection' should endanger the nation.

FLQ October crisis – need for domestic powers and the Dare report

As the Cold War continued to create international tension, with acute periods such as the Cuban Missile Crisis in 1962, domestic terrorism also occurred in countries around the world. Canada had its own experience in the 1960s with the rise of the FLQ in Quebec. Bombings, murder and other civil unrest culminated in October 1970 with the kidnapping of James Cross, the British Trade Commissioner, and of Quebec Labour Minister Pierre Laporte. These crimes led the federal government to enact the War Measures Act in order to suspend the Canadian Bill of Rights, provide military support and give police special powers to search and arrest without warrant or immediate charges. Prime Minister Trudeau, addressing the country on 16 October 1970, explained the reasons for taking these actions and said:

> These are strong powers and I find them as distasteful as I am sure do you. They are necessary, however, to permit the police to deal with persons who advocate or promote the violent overthrow of our democratic system. In short, I assure you that the Government recognizes its grave responsibilities in interfering in certain cases with civil liberties, and that it remains answerable to the people of Canada for its actions. The Government will revoke this proclamation as soon as possible.[18]

Prime Minister Trudeau went on to explain that the government was going to implement these powers cautiously and as a last resort. He also recognised the need for legislation that would be more suitable for domestic situations.

Following the October crisis a study group led by General Dare prepared a report calling for a new approach to domestic emergency management and for appropriate legislation.[19] Such legislation, in the form of the Emergencies Act 1988,[20] took another 18 years of discussion and debate before passing through Parliament.

Y2 K, 9/11: OCIPEP to PSC

The quarter century since the adoption of the Emergencies Act has been no less tumultuous. While the Emergencies Act 1988 has remained essentially untouched, its accompanying Emergency Preparedness Act 1985,[21] which set out the Federal government's emergency management roles and responsibilities, was amended in 1995,[22] and finally repealed with the passing of the Emergency Management Act in 2007.[23] The changes in how the federal government plans for emergencies have been driven by the kinds of threats the nation has faced.

The first major change was a renewed concern over critical infrastructure. While the concept of critical infrastructure, in the form of protected and controlled places and areas, goes back at least to the Defence of Canada Regulations 1939, it was the more recent concern for cyber-security that led to changes. This risk was highlighted by the Y2 K problem and various denial of service and other hacker attacks on government computer systems. As a result of this the federal government created the Office of Critical Infrastructure and Emergency Preparedness (OCIPEP) in 2000.

A little over a year later the tragic events of 9/11 set off a chain reaction of legislative changes. The Public Safety Act, 2004 amended various other acts to give government ministers more powers to create interim orders for their areas of responsibility.[24] These interim powers are crucial to the likelihood that the Governor in Council will ever see a declaration of a state of national emergency as necessary. Section 3 of the Emergencies Act 1988 defines a 'national emergency' as:

an urgent and critical situation of a temporary nature that

(a) seriously endangers the lives, health or safety of Canadians and is of such proportions or nature as to exceed the capacity or authority of a province to deal with it, or

(b) seriously threatens the ability of the Government of Canada to preserve the sovereignty, security and territorial integrity of Canada

and that cannot be effectively dealt with under any other law of Canada.

Therefore granting ministers the power to make interim orders under other laws greatly increases the range of government's alternatives to declaring a state of emergency. For example, the Food and Drugs Act 1985 was amended in section 30.1(1) so that 'The Minister may make an interim order that contains any provision that may be contained in a regulation made under this Act if the Minister believes that immediate action is required to deal with a significant risk, direct or indirect, to health, safety or the environment'.[25] This, and similar additions to other acts, allows a cabinet minister to urgently exercise powers that would normally need to go through the process of creating or amending a regulation. Of course, these orders and regulations would be derived from acts that, unless specified, would have to conform to the Canadian Charter of Rights and Freedoms, thus preserving the protection of civil rights.

The most recent changes to the federal system resulted in the assimilation of emergency management into Canada's security apparatus. The Department of Public Safety and Emergency Preparedness Act 2005 merged the remnants of OCIPEP with the Solicitor General into a new department in a way similar to that used by the Department of Homeland Security to swallow up the Federal Emergency Management Agency in the US. However, unlike in the US example, the Canadian solution did not leave emergency management as a separate entity alongside the other agencies of the former Solicitor General (the Royal Canadian Mounted Police, the Canadian Security and Intelligence Service, the Canada Border Services Agency, the Canadian Firearms Centre and the Correctional Service of Canada). This absorption of the emergency management function into the general management of the department, later to be renamed simply Public Safety Canada, was completed with the Emergency Management Act 2007.

This brief history sets the context in which the current legislation was drafted. However, the purpose of this paper is neither to judge these events nor to wade into the broader issues of what it takes to be, or govern, a sovereign nation. Perhaps the circumstances did justify the actions of the government, however extreme and unfair they appear in hindsight. Or maybe the context of the times – for example, Canada did not have universal suffrage devoid of limits based on gender or ethnicity until aboriginal Canadians were given the vote in 1960 – can explain the decisions based on ancestry and nationality during the World Wars. It is reasonable to assume that contemporary world affairs and the acute experiences of other nations may have amplified the perception of risk for decision-makers during the FLQ crisis. What is important is the influence these events had on the drafting and adoption of the Federal Emergencies Act in 1988, the creation of the various Provincial Acts over the preceding decades and the powers this diverse legislation grants to limit civil rights in times of disaster.

Emergencies Act 1988

In the best light Canada's experience with the War Measures Act was a series of decisions, taken in the heat of crises, to impose extraordinary measures deemed unavoidable under the circumstances. At worst it represented the mistreatment of residents, Canadians and foreigners living peacefully in Canada, and a disregard for civil rights. These abuses remained a point of shame that echoed in Parliament with the passing of the Emergencies Act in 1988, when the Minister of National Defence, Perrin Beatty, stated that:

> [with] the passage of this Bill from the House of Commons, we will see the abolition of the War Measures Act in Canada. Never again will we have a situation, as was the case in World War II, where Canadian citizens were interned on the basis of their racial ancestry. What we will have is a situation where there will be guarantees. Never again, for example, will there be an ability to use the War Measures Act to knock on the door in the dark of night, to sweep up our citizens, to hold them without charge and without the right of habeas corpus.[26]

Consequently the Emergencies Act 1988 acknowledges the importance of protecting civil rights in its preamble and, the more importantly, in the body of the Act.

The Preamble of the Emergencies Act 1988 sets out that the intention of the Act is to provide special temporary powers in times of national emergency. It confirms that the need for such powers must be great enough to warrant their use and that Parliament will retain oversight of how the powers are exercised. The Preamble concludes with the following statement:

> AND WHEREAS the Governor in Council, in taking such special temporary measures, would be subject to the *Canadian Charter of Rights and Freedoms* and the *Canadian Bill of Rights* and must have regard to the *International Covenant on Civil and Political Rights*, particularly with respect to those fundamental rights that are not to be limited or abridged even in a national emergency.

While these protections have been criticised for being outside the body of the Act,[27] it is clear that the government understood that the Act needed to address the concerns arising from the past applications of the War Measures Act. Following the first reading, in response to concerns raised by the National Association of Japanese Canadians, the bill was amended to read:

> s4. Nothing in this Act shall be construed or applied so as to confer on the Governor in Council the power to make orders or regulations

(a) altering the provisions of this Act; or
(b) providing for the detention, imprisonment or internment of Canadian citizens or permanent residents within the meaning of subsection 2(1) of the Immigration and Refugee Protection Act on the basis of race, national or ethnic origin, colour, religion, sex, age or mental or physical disability.

The Emergencies Act sets out powers for four different types of national emergency: public welfare, public order, international and war. Each type of emergency has its own definition of the circumstances that must be occurring and each still requires that the overall threshold of a 'national emergency' (as defined above) be met. Each also has slightly differing processes for consultation with the Provinces and for the specifics required in the declaration. There are provisions for parliamentary review, oversight and approval of all national emergency declarations and means for Parliament to revoke declarations or limit the powers being exercised.

Once a national emergency of one of the four types has been declared, the federal government, through the Governor in Council, has the power to make orders and regulations. The scope of these orders varies with the type of emergency. For public welfare emergencies, essentially natural or technological disasters, the powers are limited to those commonly available to provincial and municipal governments through the various provincial emergency measures legislation to provide for the safety and well-being of the public during the response. Public order emergencies are generated by threats to national security and the powers available to the government reflect a security-oriented response.

International emergencies are harder to imagine, but historical events such as the Korean conflict and the Cuban Missile Crisis were clearly influential in the development of the legal definition. During international emergencies, which by definition affect all of Canada and are therefore clearly and solely within federal jurisdiction, the powers available to the government extend to issues of defence production, immigration and other national matters. Declaring a war emergency gives the government almost unfettered power, within the overarching protections for civil rights, to pursue any actions necessary for the war effort except to enforce conscription into the armed forces.

Review of provincial emergency powers

At the end of the Second World War, and influenced by cold war apprehensions, the provinces began implementing their own emergency measures legislation. The Province of Manitoba, for example, passed the Emergency Measures Act in 1951 and then its replacement, the Civil Defence Act, in 1952.[28] This legislation allowed for civil disasters or war emergencies to be declared and for special powers to be granted. The recognition of the differences and similarities between civil disasters, such as floods, and war emergencies was a significant step in the development of emergency legislation. Manitoba's Civil Defence Act 1952 (and the subsequent Emergency Measures Act 1970) gave wider powers to the government during war emergencies than civil disasters, although later, when the new Manitoba Emergency Measures Act 1987 was passed, the concept of a war emergency was dropped and most of the war emergency powers were consolidated under a more generic declaration of a state of emergency.

Across Canada the provinces were going through similar processes from the 1950s onwards. As a result the emergency management legislative landscape is fairly monotonous, basically the same scenery with a few distinct regional differences, much like Canada's physical landscape. However, these acts contain the ability to grant significant powers to suspend or otherwise affect civil rights during declared emergencies and should be examined more closely. The easiest way to approach such a review is to consider the roughly two dozen powers most commonly shared by the legislation. These powers can be grouped into:

1. Authorise necessary action;
2. Control location of people;
3. Use property and control resources;
4. Require service;
5. Powers to search, arrest and fine;
6. Miscellaneous.

Appendix 1 contains a table of these powers and indicates the basic section number of the provincial and territorial legislation where they can be found. The intention of the table is to demonstrate the consistency of the emergency powers granted by federal and provincial legislation across the country. The powers in the provincial and territorial legislation are assigned to the relevant government minister or the local authority or both. Each province has a column for the powers available during provincially or minister-declared emergencies and one for local emergencies declared by mayors or local authorities. In some provinces these powers are identical, but in most the government retains certain powers not available to local officials. The powers are paraphrased in this discussion to indicate the kind of power being granted and a certain amount of judgment was exercised to assign specific powers to a particular category. The six categories indicated above will now be explored further.

Authorise necessary action

Take any action considered necessary

Almost every piece of legislation begins the section granting emergency powers with a statement allowing the government to take any action considered necessary to ensure the safety of the population and to resolve the emergency situation. This carte blanche approach is usually followed by a reference to specific powers but without attempting to limit the range of actions available to the government. One interesting alternative to this is Ontario's Emergency Management and Civil Protection Act1990s 4(1):

> The head of council of a municipality may declare that an emergency exists in the municipality or in any part thereof and may take such action and make such orders as he or she considers necessary and are not contrary to law to implement the emergency plan of the municipality and to protect property and the health, safety and welfare of the inhabitants of the emergency area.

This formulation does limit the powers available to those that are otherwise legally available to the local authority. This contrasts with other provinces' legislation, which retains the idea of temporarily exercising extraordinary powers in times of crisis.

Authorise others' responsibilities or actions

Frequently associated with the power to take any action is the power to authorise or require a person or another jurisdiction to take action. This usually means giving the provincial government the power to require a local authority to implement its emergency plan but may also be creating an ability to delegate emergency powers. Related to this is the power *to make regulations or bylaws during emergency*, which is often restricted to the province.

Activate or implement emergency plans

Almost all the provinces give the power to implement emergency plans when a state of emergency is declared. This may seem obvious but since the emergency plans may contain extraordinary measures it is appropriate to include this as an emergency power. It is important to note, however, that in some jurisdictions, such as Manitoba, emergency plans may be implemented without requiring a declaration of emergency.

Control Location of People

Regulation or prohibition of travel/roads

This is a very common power and one that may appear innocuous. Limiting access to an area during an emergency, including requiring people to obtain travel passes or otherwise control movement, is a useful power. However, Canadians are not accustomed to having to show identification or face any other restriction when traveling (other than at airports) and the right to move within the country is protected by the Canadian Charter of Rights and Freedoms. Sections 30(1)(g) and 30(1)(h) of the federal Emergencies Act 1988, under an international emergency, extend this power to include preventing Canadians from travelling outside the country and to permit the *deportation* of foreigners from Canada. The Emergencies Act also retains the power to designate and secure protected places in order to prevent trespassing or sabotage at key facilities. These powers can easily be traced back to their civil defence roots in the Defence of Canada Regulations 1939.

Evacuation of persons/removal of personal property

The flip-side of the power to prevent people from travelling into an area is the power to require people to leave the area they are already in. Forcing people to evacuate their homes or their whole community, and to remove personal property from an area, is another common response strategy that has serious implications for civil rights. The provincial legislation usually grants these powers together, although Quebec separates them, and most combine them with an obligation to provide for the *care of evacuated people or property*. In a similar vein Manitoba has a separate power for the control of movement of people and livestock related to containing the spread of disease. Very recently Manitoba also introduced the power to apprehend someone who has not complied with an evacuation order and to use reasonable force, including entry without a warrant, to remove that person to a place of safety.[29]

Close public or private places/prevent entry

Also connected to the power to control people's location during an emergency, some provinces (Ontario, Quebec, Prince Edward Island, Newfoundland and Labrador, and Nunavut) create the opportunity for the government to *close public or private places during emergencies or to prevent entry* into identified buildings. Quebec and Newfoundland also have a related power, under certain circumstances, to order curfews or confinement. The Federal Emergencies Act 1988 also allows the government to regulate public gatherings during a public order national emergency.

Use property and control resources

Requisition, use or disposition of property

All jurisdictions in Canada provide some provision during a declared emergency for the government to requisition, use or dispose of property. Such ability to commandeer property varies in the precise wording. This is another set of powers that has been carried forward from the Second World War for use in peacetime emergencies.[30] One variation of this power to use property is the ability to order the *demolition of property*. British Columbia's provision states that the ministers may 'cause the demolition or removal of any trees, structures or crops if the demolition or removal is considered by the minister to be necessary or appropriate in order to prevent, respond to or alleviate the effects of an emergency or disaster'.[31] This is fairly indicative of the other provinces' legislation.

Regulate essential resources

The other major power that transferred from the Defence of Canada Regulations 1939 into the provincial legislation for natural disasters is the ability to *regulate essential resources* and, under some circumstances in a handful of jurisdictions, to take over the control and care of utilities, services and industries deemed critical to the response. There is often a specific reference, either as a separate power or combined with the more general power to maintain services, to provincial governments' and municipalities' ability to establish shelters, hospitals and welfare centres.

Amendments in 2013 to Manitoba's Emergency Measures Act 1987 introduced the concept of 'critical service providers' and the requirement that companies so designated, with telecommunication companies as likely candidates, prepare business continuity plans.[32] This may signal a change to the approach of giving the government the power to repair or maintain services and place the obligation more squarely on the providers to be prepared.

With the power to control essential resources and services comes the power, in some provinces, to *implement price fixing*. The wording varies and the power may be linked to the control of essential resources or provided for as a separate power. Ontario allows for 'fixing prices for necessary goods, services and resources and prohibiting charging unconscionable prices in respect of necessary goods, services and resources.'[33] In the unique case of an international emergency, declared under the federal Emergencies Act 1988, this power relates to the control and investigation of black-marketeering.

Require service

Direction to render qualified service

In parallel to the powers to control and use property or regulate the delivery of services are the powers to engage human resources. All the current legislation across Canada, with the exception of the Yukon,[34] contains explicit powers for the government to have qualified persons or classes of persons render such services as they are qualified to provide. Ontario only allows for the 'authorisation' of such services but stops short of the power many other provinces wield to require such service. Most jurisdictions also have sections in their emergency legislation regarding *compensation* connected to the requirement to provide service and more generally to losses incurred as the result of the use of property.

More contentious, perhaps, is the power of *conscription of persons*. This power exists in Alberta, Saskatchewan, Nova Scotia, New Brunswick, Prince Edward Island (PEI), the Northwest Territory and Nunavut. This can be expressed simply, as is the case in Saskatchewan, as 'conscript persons needed to meet an emergency',[35] or with some reference to compensation, such as in PEI, where the minister may 'order the assistance, with or without remuneration of persons needed to carry out the provisions mentioned in [the section listing emergency powers].'[36]

Since the intention of emergency legislation is to grant special powers that are not otherwise available to the government, it is important to note that there may be other legislation granting similar powers. For example, while Manitoba's Emergency Measures Act 1987 does not allow for conscription, its Wildfires Act states: 'An officer may order any able-bodied adult located in the vicinity of a wildfire to assist in wildfire protection operations when sufficient numbers of firefighters or volunteers are not immediately available.'[37] Therefore this power to conscript may exist under other circumstances in some jurisdictions.

This ability of a government to require members of the public to provide services, either relating to a qualification or simply through conscription and with or without compensation, presents significant civil rights concerns. The United Nations' International Covenant on

Civil and Political Rights states: 'No one shall be required to perform forced or compulsory labour',[38] although this is qualified later: 'For the purpose of this paragraph the term "forced or compulsory labour" shall not include: (iii) Any service exacted in cases of emergency or calamity threatening the life or well-being of the community.'[39] So the power available through the emergency legislation is valid but should still be exercised with caution.

Powers to search, arrest and fine

Entry and search

The final cohesive set of powers relates to actions the public could normally expect to be exercised by the police. However, under a declared state of emergency, who exercises these powers and the process required to take these actions may be very different. Only Ontario and the Yukon do not have the clear power to *enter and search* a property during an emergency. The other provinces provide an option to authorise their agents, not necessarily limited to sworn peace or police officers, to enter and search premises without a warrant. In British Columbia this is phrased as 'authorize the entry into any building or on any land, without warrant, by any person in the course of implementing an emergency plan or program or if otherwise considered by the minister to be necessary to prevent, respond to or alleviate the effects of an emergency or disaster'.[40]

Imposing fines is another aspect of the emergency legislation which could be seen as a police function. Only the Northwest Territories' Civil Emergency Measures Act 1988 does not contain provisions for what is considered an offence and for assigning imprisonment or fines to those convicted of committing an offence. The fines vary across the country and some legislation differentiates between individuals and corporations.

The power to arrest or detain individuals had been absent from the legislation since the end of the Defence of Canada Regulations but was recently revived in Manitoba. A recent amendment there now allows for a peace officer to arrest someone committing an offence in order to establish their identity, collect evidence or prevent repeating offences under the Act.[41]

Miscellaneous powers

The remaining powers found scattered across the country's emergency legislation cover a range of topics. These may be a reflection of past events when a particular issue was identified and a power then established or they may be left-overs from earlier versions of the legislation. Most provinces provide some authorisation to make emergency payments or otherwise allow government to commit resources during an emergency. Several have powers to construct works or make repairs and Nova Scotia echoes the federal Emergencies Act in allowing the government to assess damage to works or to the environment.

Conclusion

Canada has rightly enacted laws at the federal and provincial level to allow the government to take extraordinary measures in times of emergencies. In doing so the provisions of Article 4 (1) of the International Covenant on Civil and Political Rights are met, specifically:

> In time of public emergency which threatens the life of the nation and the existence of which is officially proclaimed, the States Parties to the present Covenant may take measures derogating from their obligations under the present Covenant to the extent strictly required by the

> exigencies of the situation, provided that such measures are not inconsistent with their other obligations under international law and do not involve discrimination solely on the ground of race, colour, sex, language, religion or social origin.

Canada has created the required legal framework to derogate from the Covenant and permit certain limitations on civil rights when and where necessary and only for the time needed. This is clear in the preamble of the federal Emergencies Act.

Provincially, Ontario's legislation is unique in its explicit protection of the civil rights during an emergency:

> 7.0.2 (1) The purpose of making orders under this [Emergency Powers and Orders] section is to promote the public good by protecting the health, safety and welfare of the people of Ontario in times of declared emergencies in a manner that is subject to the Canadian Charter of Rights and Freedoms.

No other province makes this point as clearly. In this way the Ontario legislation, derived from its own unique set of experiences,[42] is more in line with the federal government's commitment to only using special temporary powers in times of urgent need and in the absence of alternatives. The lack of language in the other provincial legislation to guarantee protection of civil rights is worrisome.

The history of civil rights abuses in Canada during the two World Wars and the FLQ crisis under the War Measures Act may seem increasingly irrelevant as time fades the memories. However, the quarter century since the introduction of the Emergencies Act 1988 does not necessarily demonstrate its success, given the complete lack of any application of that Act. Furthermore, the powers that are regularly exercised across Canada are authorised by the provincial legislation and the majority of these powers originate from a civil defence approach to war emergencies, even though they are now used for peacetime disasters.

Thus, the legislation in Canada is heavily weighted towards providing extraordinary powers for disaster response but does not have a balanced emphasis on hazard mitigation. Donald Geis argues that living in what he coined a 'disaster resistant community' – 'the safest possible community that we have the knowledge to design and build in a natural hazard context' is a basic human right.[43] If this proposition is accepted,[44] if it can be agreed that citizens should expect their governments to ensure their safety as a fundamental factor in community decision-making, then citizens should further question the need for all these legislative powers to suspend civil rights in order to overcome events that often could have been avoided or mitigated in the first place. At a minimum, citizens should be protected before an emergency by an equally potent legislative approach to risk reduction. Then, at least, the application of powers to suspend civil rights during an emergency could be seen as a true last resort, rather than as a convenience, and the powers could be further restricted to guarantee rights and liberties.

Notes

1. S.C. 1988, c. 29.
2. Public Safety Canada, 'Canadian Disaster Database', http://www.publicsafety.gc.ca/cnt/rsrcs/cndn-dsstr-dtbs/index-eng.aspx (accessed October 29, 2013).
3. Donald Geis, 'By Design: The Disaster Resistant and Quality-of-life Community', *Natural Hazards Review* 1, no. 3 (2000): 151.
4. The Constitution Act, 1982, being Schedule B to the Canada Act 1982 (UK), 1982, c 11.

5. Herbert Marx, 'Emergency Power and Civil Liberties in Canada', *McGill Law Journal* 16 (1970): 39, 56.
6. W.P.M. Kennedy, 'The Interpretation of the British North America Act', *Cambridge Law Journal* 8, no. 2 (1943): 146, 147.
7. The Constitution Act, 1982 did not amend the Constitution Act 1867 in this respect. Schedule item 1.
8. 5 George V Ch 2 1914.
9. *Canada Gazette* xlviii (28 October 1914): 1304.
10. *Canada Gazette* xlviii (30 October 1914): 1382.
11. Registration of Alien Enemies 28 October 1914 s 9, *Canada Gazette*, xlviii : 1304 (emphasis added).
12. Bohdan Kordan, *Enemy Aliens, Prisoners of War: Internment in Canada during the Great War* (Montreal: McGill-Queen's University Press, 2002).
13. Elliot Tenofsky, 'The War Measures and Emergencies Acts: Implications for Canadian Civil Rights and Liberties', *American Review of Canadian Studies* 19, no. 3 (1989): 293.
14. Lester Phillips, 'Canada's Internal Security', *Canadian Journal of Economics and Political Science* 12, no. 1 (1946): 18.
15. David McConnell, *Plan for Tomorrow...TODAY! The Story of Emergency Preparedness Canada 1948–1998* (Ottawa: Heritage Research Associates, 1998).
16. Phillips, 'Canada's Internal Security'.
17. Prime Minister St-Laurent, *Hansard*, February 20, 1951: 498.
18. http://www.collectionscanada.gc.ca/primeministers/h4-4065-e.html (accessed December 21, 2013).
19. McConnell, *Plan for Tomorrow...TODAY!*.
20. R.S.C., 1988, c. 22.
21. R.S.C., 1985, c. 6. The Emergency Preparedness Act established Emergency Preparedness Canada and set out the requirements for emergency planning, whereas the Emergencies Act only provides powers and processes related to declaring a State of National Emergency.
22. Bill C-65 (an act to reorganise and dissolve certain federal agencies) received Royal Assent on 13 July 1995 and resulted in Emergency Preparedness Canada being returned to the Department of National Defence.
23. S.C. 2007, c 15.
24. S.C. 2004, c 15.
25. R.S.C., 1985, c. F-27.
26. *Hansard*, July 11, 1988, as quoted in 'Historic Day for Canada – War Measures Act Repealed', *Emergency Preparedness Digest*, July–Sept. (1988): 34.
27. Peter Rosenthal, 'The New Emergencies Act: Four Times the War Measures Act', *Manitoba Law Journal*, 20 (1991): 563.
28. Emergency Measures Act S.M. 1951 c.18; and Civil Defence Act S.M. 1952 (1st Sess.), c. 8, s. 1.
29. Emergency Measures Act 1987 s 18(1).
30. Defence of Canada Regulations 1939 s 48.
31. Emergency Program Act RSBC 1996 c 111 s 0(1)(j).
32. Emergency Measures Amendment Act 2013 CCSM c.12.
33. Emergency Management and Civil Protection Act RSO 1990 c E.9 s 7.0.2(4)(11).
34. Civil Emergency Measures Act RSY 2002 c 34.
35. Emergency Planning Act SS 1989–90 c E-8.1 s 18(1)(m).
36. Emergency Measures Act RSPEI 1988, c E-6.1 s 11(k).
37. Wildfires Act CCSM c W128 s 7(12).
38. United Nations' International Covenant on Civil and Political Rights, Article 8 s 3(a).
39. Ibid.,, Article 8 s 3(c).
40. Emergency Program Act RSBC 1996 c 111 s10(1)(i).
41. This power, created in s17(2) of the Emergency Measures Amendment Act 2013 CCSM c12, is aimed at preventing people from interfering with emergency measures, such as the operation of flood control works.
42. Daniel Henstra, 'The Dynamics of Policy Change: A Longitudinal Analysis of Emergency Management in Ontario, 1950–2010', *Journal of Policy History* 23, no. 3 (2011): 399.
43. Geis, 'By Design', 151, 153.

44. There is some support from the European Court of Human Rights: *López Ostra v. Spain*, App no 16798/90, Series A no. 303-C (1994); *McCann v UK*, App. no.18984/91, Ser A vol.324 (1995); *Budayeva and Others v. Russia*, App nos. 15339/02, 21166/02, 20058/02, 11673/02 and 15343/02, 20 March 2008; *Murillo Saldias and Others v. Spain*, App no. 76973/01, 28 November 2006; and *Kolyadenko and others v. Russia*, App nos. 17423/05, 20534/05, 20678/05, 23263/05, 24283/05 and 35673/05, 28 February 2012.

Appendix 1 Emergency Powers in Canadian Legislation

Power	Canada					BC		AB		SK		MB		ON		QC		NS		NB		PE		NL		NWT		Yukon		NU	
	DC Rgs	PW	PO	In	War	P	M	P	M	P	M	P	M	P	M	P	M	P	M	P	M	P	M	P	M	T	M	T	M	T	M
take any action considered necessary					40	10	13	19	24	18	21	12	12	7.0.2	4			14	14	12	12	11	11	11		12	17	9	9	13	
authorise others' responsibilities or actions				30		10	13	19	24	18						93	47			12		11				12	17	9		13	18
activate or implement emergency plan						10	13	19	24	18	21	12	12	7.0.2		93		14	14	12	12	11	11			12	17	8	8	13	18
make regulations /bylaws during emergency						28		19		18																22		9	9		
regulation or prohibition of travel/roads	20	8	19	30		10	13	19	24	18	21	12	12	7.0.2		93	47	14	14	12	12	11	11	11	405	12		9	9	13	
deportation				30																											
close public or private place/prevent entry														7.0.2		93						11	11		405					13	
evacuation of persons	32	8				10	13	19	24	18	2	12	12	7.0.2		93	47	14	14	12	12	11	11	11	405	12				13	
removal of personal property		8				10	13	19	24	18	2	12	12	7.0.2		93		14	14	12	12	11	11	11		12				13	
care of evacuated people or property		8				10	13	19	24	18	21	12	12	7.0.2		93	47	14	14	12	12	11	11	11		12				13	
control movement of people/livestock with disease												12	12																		
entry and search of any dwelling-house, premises, conveyance or place	58, 59			30		10	13	19	24	18	21	12	12			93		14, 22	14, 22	12, 23	12, 23	11, 18	11, 18	11		12				28	28
requisition, use or disposition of property	48	8	19	30		10	13	19	24	18	21	12	12)	7.0.2				14	14	12	12	11	11	11		12	17		9	13	18

Power	Canada					BC		AB		SK		MB		ON		QC		NS		NB		PE		NL		NWT		Yukon		NU	
demolition of structures, trees, crops						10	13	19	24	18	21	12	12	7.0.2		93		14	14	12	12	11	11	11		12	17			13	18
designation and securing of protected places	3		19	30																											
direction to render qualified service		8	19	30		10	13	19	24	18)	21	12	12	7.0.2		93	47	14	14	12	12	11	11	11		12	17			13)	18
conscription of persons needed								19	24	18	21							14	14	12	12	11	11			12				13	
compensate for service/damage		8	19			19		19	24	18	21	13				94	48									12	17			24	25
regulate essential resources		8				10	13	19	24	18	21	12	12	7.0.2		93		14	14	12	12	11	11		405	12				13	
control and care of utilities/services/ industry	13		19	30		10	13	19	24	18		12)	12	7.0.2		93		14	14	12	12	11	11	11				9	9	13	
implement price fixing				30		10	13	19	24	18				7.0.2				16		12	12			23		12				13	
authorize emergency payments		8		30		16	13		24		21	12	12			93	47	14	14												18
establish shelters, hospitals, welfare		8				10	13	19	24	18	21			7.0.2		93	47	14	14	12	12	11	11	11		12		9	9	13	
assess damage to works		8																14	14												
assess damage to environment		8																14	14												
construct works or make repairs		8				10	13							7.0.2		93		14	14											13	18
impose fines	63	8	19	30		27		17		16		20		7.0.1		128		23		24		19		22				11		33	
permit censorship	13, 14, 39																														
arrest or detain	24, 60											18, 20																			
order curfews/ confinement	36															93	47								405						
manage waste														7.0.2																	
manage information														7.0.2																	
assist in law enforcement																												9	9		

Power	Canada					BC		AB		SK		MB		ON		QC		NS		NB		PE		NL		NWT		Yukon		NU	
assist in fire fighting																												9			
protect health and welfare																												9	9		
regulate public gatherings			19																						405						
regulation of international financial activities				30																											

Note: DC Rgs = Regulations for the Defence of Canada 1939; PW = Public welfare; PO = Public order; IN = International; P = Provincial; M = Municipal; T = Territory. Provincial Legislation is as follows: BC = Emergency Program Act, RSBC 1996, c 111; AB = Emergency Management Act, RSA 2000, c E-6.8; SK = Emergency Planning Act, SS 1989-90, c E-8.1; MB = Emergency Measures Act 1987 C.C.S.M. c. E80; ON = Emergency Management and Civil Protection Act, RSO 1990, c E.9; QB = Civil Protection Act, RSQ, c S-2.3; NB = Emergency Measures Act, RSNB 2011, c 147; NS = Emergency Management Act, SNS 1990, c 8; PE = Emergency Measures Act, RSPEI 1988, c E-6.1; NL (P) = Emergency Services Act, SNL 2008, c E-9.1; NL(M) = Municipalities Act SNL 1999 c M-24.

'Keep Calm and Carry On': informing the public under the Civil Contingencies Act 2004

Rebecca Moosavian

School of Law, Northumbria University, Newcastle, UK

Part Iof the Civil Contingencies Act (CCA) 2004 provides a framework that governs the planning and preparations for a wide range of potential emergencies. It requires the engagement and cooperation of numerous 'public' bodies, including central government, local authorities and the emergency services, as well as a range of 'private' organisations such as utilities companies. It is apparent that information plays a fundamental role in the Part I planning provisions and associated guidance. This article will focus on one specific information-related provision, namely the duty to maintain plans to warn and inform the public in the event of an emergency. It undertakes detailed analysis of the CCA 2004 provisions, secondary legislation and extensive government guidance regarding informing the public and related media-handling. It analyses these measures in light of two competing organisational models identified by Walker and Broderick as being present across various CCA 2004 arrangements. The tensions between traditional authoritarian 'command' structures on the one hand, and more flexible, decentralised arrangements, on the other, are particularly pronounced in the context of informing the public. They reflect the challenges of maintaining political authority while handling and disseminating information that is fluid in nature and evades control.

Introduction

Part I of the Civil Contingencies Act 2004 (CCA 2004) provides a framework that governs the planning and preparation for a wide range of post-cold war potential emergencies, including terrorist incidents, cyber-attacks and natural hazards such as flooding or extreme weather.[1] The sturdy, reliable planning required by Part I has been relatively neglected by lawyers, in favour of the potential dystopian dangers of government emergency powers in Part II. Nevertheless, emergency planning is a global concern that fuels a vast interdisciplinary research industry,[2] and further scrutiny of Part I reveals illuminating material regarding modern government in the 'Information Age'.

Across various accounts of emergency planning literature there emerges a basic, recurring tension between two broad organisational structures. The first model, characterised as 'command-and-control', is a centralised hierarchy directed by a clear single point of leadership. In contrast, the second approach is decentralised, flexible and locally situated in nature. The tension between both models is especially pronounced regarding the function and organisation of information within the CCA 2004 regime. The role of information dominates Part I, arguably reflecting the Pitt Review's claim that it is 'the lifeblood of effective emergency planning'.[3] However, this article will examine just one specific information-related measure in the CCA 2004, its regulations,[4] and associated government guidance: provisions to warn and inform the public in the event of an emergency.

The article will start by providing a brief outline of the emergency planning regime, before providing an overview of the two alternative models. It will then look at how these models operate in the context of warning and informing the public under the CCA 2004.

Emergency planning: overview

To achieve local-level emergency planning across the country the CCA 2004 places legal obligations on a range of relevant organisations, divided into two groups. The first, termed 'Category 1 responders' because they form the front-line of local emergency services, includes local authorities, the police, fire brigade and ambulance services.[5] The second, termed 'Category 2 responders', may also be heavily involved in specific types of emergency and includes private organisations such as utilities companies and transport operators.[6] Although this latter group's involvement in various aspects of emergency planning is required under the CCA 2004, its duties are less onerous than those of 'Category 1s', who bear primary responsibility for such planning.[7] Within each locality (based on police districts) these Category 1 and 2 responder groups are required to cooperate, share relevant information and participate in a local group called the 'Local Resilience Forum' (LRF).[8] This forum organises and performs various CCA 2004-prescribed emergency planning activities, although it has three main related tasks, all of which are undertaken by its category 1 members. First, drawing on government guidance[9] and local expertise,[10] the LRF carries out risk assessments of potential emergencies that may affect the locality.[11] Second, the LRF compiles a 'Community Risk Register' based on the risk assessments.[12] This register sets out the various potential emergencies, their likelihood and potential impact. Third, the LRF produces a range of emergency plans which outline the response arrangements to be followed if particular emergencies occur.[13] In the event of an emergency local responders use the plans to guide their response. Where an emergency is more serious, select senior members of the LRF will form a 'Strategic Co-ordinating Group' (often termed 'Gold Command').[14] This group, usually chaired by a senior police representative, provides management and strategic leadership to direct the emergency response.[15]

The CCA 2004 supporting regulations and extensive government guidance detail how these arrangements should be put into practice by prescribing matters such as: the methods of planning; the format of documents; the procedures for allocating planning tasks to responders and information-sharing etc; standards of 'good practice' and model protocols.[16] Although planning occurs at a local level, this is overseen by central government; all responders are accountable to ministers for their emergency planning activities,[17] and must take account of central government guidance in their planning,[18] such as the National Risk Register for Civil Emergencies.[19] Central government will also become involved where an incident escalates; in a 'significant' level-1 emergency it will provide the leadership of an appointed 'Lead Government Department' with expertise in the relevant emergency area.[20] Alternatively 'serious' level-2 emergencies and above will involve the leadership of the Cabinet Office Briefing Rooms (COBR) committee of senior ministers.[21]

Command v. decentralised structures

The influence of two alternative organisational models, one hierarchical and the other decentralised, is apparent across various parts of the CCA 2004. For example, when discussing potential forms of relationship between responder agencies, Walker and Broderick distinguish between 'a purely hierarchical system, in which all is directed or at least cascaded from the centre' and, alternatively, 'a networked or nodal model, whereby emergency planners coalesce in horizontal, heterarchical partnerships in the co-production of security',[22] the latter being generally supported as more effective for emergency planning purposes. This part outlines the relevant properties and theoretical justifications associated with both models.[23]

'Command-and-control' model

The first model, characterised as 'command-and-control', assumes authority vested in a single individual (or group) who can provide strong, decisive leadership. It therefore involves a distinct hierarchical, top-down linear structure where a 'commander' 'controls' the 'commanded'. This model, based on a somewhat authoritarian leadership culture, has been particularly cogent in times of crisis or war.[24] Its influence, therefore, in emergency planning,[25] and response, at national and local level is to be expected. The rationale for this model is exemplified by Garth Whitty, who claims:

> Emergencies most frequently develop into disasters because of the inadequacies of command and control. Confidence in commanders is established before an incident occurs and is dependent upon a number of qualities, the most important of which is leadership.[26]

Similarly, noting the importance of strong leadership in emergency response, though in more circumspect terms, John Twigg refers to 'the history, character and culture of disaster work, with its command and control mentality, blue-print planning, technocratic bias and disregard for vulnerable communities' knowledge and expertise'.[27]

Command-and-control features are apparent across the CCA 2004 regime. The model's influence is particularly marked in Part II of the Act, which confers emergency powers on ministers, but also in Part I regulations that allow ministers to issue orders or urgent directions that category 1 responders must perform.[28] Such tendencies also influence central government's ongoing desire to maintain its traditional monopoly regarding national security-related information.[29] Further hegemonic tendencies are evident in the Act's disparate allocation of legal duties between central government and local responders; this pattern results in 'bottom-heavy' responsibilities and a lack of higher accountability.[30] Yet, as guidance indicates, there remains 'a public expectation that [government] will give a strong lead at a time of crisis, take charge of events and manage situations'.[31]

The centrality of the command-and-control model in emergency response closely reflects traditional demands for a strong protective leader in times of crisis or disaster. This model is very much epitomised by Thomas Hobbes' *Leviathan*, written during the chaos of civil war.[32] That background climate clearly influenced Hobbes' account of the lawless 'state of nature' as chaotic, disordered and uncertain, filled with conflict, 'continuall feare and danger of violent death',[33] a hypothetical disaster zone of sorts. In response, Hobbes' social contract advocates the creation of a strong, powerful sovereign monarch whose command can maintain peace, order and individual security.[34] In doing so, this patriarchal monarch must be given the unconditional obedience of the populace.[35] Hobbes' grand narrative legitimises modern liberal nation-states, yet it is also, as critics have suggested, a model ultimately grounded in fear.[36] Such dynamics are arguably implicit in modern approaches to emergency management, where government policies are buttressed by rhetoric that, for example, terms the present era the 'age of uncertainty'.[37] Yet, as modern theorists explain, this uncertainty is 'manufactured'; a paradoxical and inescapable 'consequence of scientific and political efforts to control or manage (earlier) [risks]'.[38] That the four main threats identified in the UK's *National Security Strategy* are primarily human-made tends to support this analysis.

Decentralised model

The 'decentralised' model in emergency planning represents contrasting characteristics to the command model. This form of arrangement is less formally structured and more flexible. It does not situate authority at a single point issuing top-down commands, but is instead non-hierarchical and works at local, ground level. It entails people working in a horizontal network-type arrangement in a collaborative, cooperative and democratic culture.[39] This model is most influential in the CCA 2004 regime's strong emphasis on cooperation between local responders,[40] and provisions to facilitate information-sharing.[41] Walker and Broderick identify these measures as decentralised in nature, and claim that, in making institutional arrangements to attain resilience, 'the late modern state is being *forced* to adopt styles of governance intended to achieve influence in complex horizontal (or heterarchical), decentred and networked relationships, with an emphasis on shared knowledge and consensus'.[42]

This comment indicates that the decentralised model is not necessarily compatible with the pre-existing culture or dynamic of the state in emergency management mode (which arguably more comfortably fits with the command model). It implies that this decentralised model is not the preferred or natural choice, but is nonetheless necessary for emergency management. One crucial underlying reason for this is arguably the nature of information that constitutes the very 'lifeblood' of the system.

The decentralised model is broadly reflective of, and more compatible with, the nature of information itself. With the emergence of what some commentators term the 'Information Society' rather than the 'Risk Society',[43] have come theories exploring the nature and properties of information, particularly in light of technological developments. In this context three related features of information are significant. First, information is intangible, incorporeal and cannot be understood as a physical thing.[44] Digitisation replaces physical information-storage with a 'complex – and highly liquid – pattern of ones and zeroes'.[45] This view of information, or rather its movements, as being liquid in character is also evident in common accounts of information 'flows',[46] and 'leaks'. Pitt's depiction of information as 'lifeblood' similarly draws upon a metaphor of liquidity.[47] Second, as Lash argues, information acts to compress time, as it travels distances rapidly, allowing real-time communication.[48] For Lash:

> The primary qualities of information are flow, disembededness, spatial compression, temporal compression, real-time relations. It is not exclusively, but mainly, in this sense that we live in an information age.[49]

Although rapid transmission is a clear consequence of advancements in digital technology, Lash explains that this is a continuation of trends that started with the emergence of the mass media in earlier times.[50] Third, information works in a non-linear and disjointed way, which Lash refers to as 'a quasi-anarchy of information proliferation and flows'.[51]

It works via a range of networks (at individual level, but with global reach) which play a crucial role in the Information Society.[52] Lash claims that technology stretches and breaks the linear bonds associated with the nation-state. But these bonds are 'reconstitute[d] as the links of non-linear and discontinuous networks'.[53] He also notes the decline of traditional organisations,[54] into 'disorganisations' that 'are perhaps less hierarchical than horizontal', and more fluid, mobile and reflexive.[55] Such developments raise questions about the standing of the nation-state and power more generally.[56] Because 'Information wants to be free',[57] it poses major challenges for those who wish to tightly control or monopolise it in the Information Age.

Summary

Both models reflect alternative, even opposing, properties that may be necessary for emergency planning and response. There is a long-established tradition of the need for clear, decisive leadership in an emergency. Yet, in our technologically complex 'Information Society', any information-reliant endeavour, such as emergency preparation and response, must be arranged to enable information to be handled effectively, and must therefore understand its nature. How do these conflicting models, and the values they represent, operate in the context of warning and informing the public?

Warning and informing the public

Government guidance states that 'Communicating with the public is an integral part of preparing for and responding to incidents and cannot be done in isolation';[58] and 'Good public communication is vital to the successful handling of any emergency and should be incorporated in all contingency planning'.[59] So, as part of the general duty to assess and plan for emergencies,[60] Category 1 responders must 'maintain arrangements to warn the public, and to provide information and advice to the public, if an emergency is likely to occur or has occurred'.[61] Guidance divides this duty into two specific functions. First, it requires those responders to inform the public about the risk of emergencies in their locality by undertaking ongoing awareness-raising, education and community engagement.[62] Second, it requires responders to have plans in place to warn and advise the public in case an emergency actually arises, encompassing, for example, evacuation alerts and practical advice about what action to take. Although the two functions are related, the second aspect of the duty is the focus of this article. This second aspect of warning the public must be covered by emergency plans,[63] and, in turn, responders must follow ('regard') these plans when advising the public in an emergency.[64] This ensures that a pre-prepared communications strategy, covering before, during and after an emergency, is 'fully integrated' into emergency plans.[65]

Guidance divides the public into three broad groups, each with different informational needs: first, those directly involved in the emergency; second, local people and/or relatives of those directly involved; and third, the wider public and news media.[66] Guidance also highlights the needs of vulnerable groups within the public, such as the elderly, those with mobility issues or non-English-speaking groups.[67] However, across various groups, the basic rationale for the 'public advice' duty is 'the belief that a well-informed public is better able to respond to an emergency and to minimise the impact of the emergency on the community'.[68] Better public information will enable people to make better decisions and is therefore empowering.[69]

CCA 2004 Arrangements: Overview

Under the CCA 2004 category 1 responders must cooperate to appoint one of their number as having lead responsibility for maintaining and implementing plans to inform the public in the event of an emergency.[70] Public information plans may cover general matters such as the generic advice to 'go in, stay in, tune in',[71] deployable in various emergencies, or relate to more specific emergencies, e.g. instructions to evacuate a particular area such as a city centre.[72] The lead 'must' be able to collaborate with other category 1 responders, and 'must' inform them of its actions (and proposed actions) to inform the public;[73] the latter in turn 'must' regularly consult with the lead on these matters.[74] Collaboration, partnership and information-sharing between category 1 responders is 'the critical element in the effective delivery of information to the public' and protocols can be used to help achieve this goal.[75]

Guidance sets out more detailed arrangements for informing the public in the event of an emergency, with particular emphasis on managing media interest in an emergency.[76] It confirms that responder information should be coordinated from the outset of an emergency.[77] This might entail arrangements such as a shared media briefing, a central press office and the involvement of a media liaison officer.[78] As the emergency develops, this aspect of response may be headed up by an experienced media communications specialist.[79] Where Gold Command is involved in directing an emergency response, it too must 'implement media-handling and public communications plans',[80] although this can be delegated. Gold Command will also be supported and advised by a media communications 'cell'.[81] These arrangements ensure that responsibilities are clearly allocated, preferably in advance, to enable senior personnel to provide clear leadership in informing the public.

Central government occupies a background role at this planning stage; the Cabinet Office issues national guidance on informing the public for responders to follow. The Cabinet Office is in turn advised by an independent committee whose stated objectives are to improve warning mechanisms and to ensure public information is timely and effective.[82] Where an emergency becomes national in scale, central government will become involved in managing information to the public.[83] A News Co-ordination Centre can be set up at national level.[84] Government will 'support and complement' local responders' efforts by undertaking various activities, including 'determining the public information strategy and co-ordinating public advice in consultation with [Gold Command]'.[85]

Maintaining authority in an emergency

The concern of CCA 2004 guidance in fostering clear, hierarchical leadership in emergency handling extends to communicating with the public. The reason for this is arguably the CCA 2004 regime's focus on the needs and response of the public, or specific groups within the public.[86] Overall the regulations emphasise the importance of avoiding unnecessary panic. They state that Category 1 responders '*must* have regard to the importance of not alarming the public unnecessarily' when publishing emergency plans and when maintaining arrangements to warn and advise the public in an emergency.[87] Related to this, guidance indicates a recurring concern with maintaining public confidence in an emergency. The *National Risk Register* recognises that most emergencies will have a psychological impact upon the public affected.[88] The purpose of providing information in such circumstances is for practical guidance and vital reassurance,[89] although it is widely acknowledged that panic and the breakdown of social order are 'disaster myths'.[90]

One method of fostering public calm and confidence is via a reassuring authority figure. This will be particularly vital to those directly involved in the emergency.[91] Yet guidance suggests that this should also be planned for those in the vicinity of an emergency because 'the majority of people need leadership and direction; some need to be controlled . . . the majority of individuals are likely to need some level of support'.[92] More broadly there is a concern not just to have clear command structures in place, but to ensure that leadership is also visible to the public. For example, *Emergency Preparedness* recommends that trained spokespersons will be more effective 'if...already recognised as a trustworthy authoritative person', for example 'because they already have a good public profile in the area or are in uniform (research shows that the public have great confidence in spokespeople from the emergency services).'[93] This emphasis on the uniform as a powerful visual indicator reflects guidance advising authorities to make use of images.[94] Similarly the Pitt Review into the 2007 floods across England found that, in areas where Gold Command was used effectively there was more visible leadership and better involvement of media, which helped essential information to reach the public.[95] Pitt also noted the value of local leaders with a high media profile, who provided reassurance to the public and used the media to communicate advice.[96] Thus strong leadership in emergencies is insufficient per se; it must also effectively project itself when communicating with the public. The aim is to maintain authority and the appearance of authority in order to foster public trust and reassurance.

Authoritative information in an emergency

The maintenance of responder authority is partially reliant upon its capacity to provide information that is itself authoritative: 'The flow of authoritative information . . . underpins the resilience of a community to disruptive challenges'.[97] Thus CCA 2004 guidance affords much attention to ensuring that information, particularly that provided to the public, is 'authoritative' and credible.[98] A useful starting point for such examination is the core objective of public communications in an emergency:

> When an emergency occurs, the key communications objective will be to deliver accurate, clear and timely information and advice to the public so they feel confident, safe and well informed.[99]

To foster public confidence the information it is provided must be clear, accurate and provided quickly. Rapid dissemination will be particularly essential for early alerts or warnings of sudden emergencies, which must 'reach as many people as possible as quickly as possible'.[100] But beyond this, government guidance suggests that in the first hour of an emergency the public needs information regarding the basic details of the incident, health implications, practical advice (such as what to do, where to go) and reassurance if necessary.[101] Next, clarity requires any technical information, e.g. scientific guidance regarding the health or environmental implications of a particular emergency, to be communicated in terms that people can understand.[102] But, most importantly, it requires a unified and consistent message to the public; the importance of this is repeatedly emphasised in guidance.[103] Appointing a lead responder to inform the public assists in this aim by avoiding unnecessary duplication of information and conflicting messages.[104] The ideal is responders 'speaking with one voice',[105] and this is especially important regarding emergency warning systems, as during evacuations.[106] Finally, information to the public must be accurate.[107] Factually correct information is vital to ensure the public is adequately informed about the situation so they can make informed decisions.[108] In this sense it contributes to public empowerment.[109]

Information must also be effectively disseminated, so the guidance states that: 'The key to effective communication with the public is getting the message right for the right audience'.[110] One key way in which the CCA 2004 regime seeks to ensure that the 'right message' reaches the 'right audience' is by utilising the media, particularly local radio and television, as a means of dissemination.[111] Although media organisations are not afforded Category 2 responder status, they do have a 'public service' role or 'duty' in the event of an emergency.[112] During the 2007 floods, the media, particularly local radio, undertook this role by providing information and updates about the flood.[113] Guidance confirms that in the event of an emergency the media can deliver information to large numbers of people and it is often the quickest most effective means of delivery.[114] So in emergency situations media organisations, particularly television and radio, may undertake a key 'conduit' function,[115] channelling vital messages from responders to the wider public. For this reason CCA 2004 guidance states that media organisations have a key role to play and should be encouraged to participate in LRFs and planning.[116] It also advises ongoing liaison and contact-building with media representatives so that parties can work together effectively in the event of an emergency.[117]

Summary

Overall, authority (actual or perceived) partly rests on having command of information. The ideal set out by government guidance is fast, clear authoritative information issued from a single, credible official source and disseminated widely via a conduit media or, in Twigg's terms, 'information command and control'.[118] Discussing warning systems, Sorensen claims that, although this 'linear' communications model, which 'assumes a top-down flow of 'official' warning information to the public' is 'not without merit', it 'needs to be revised' in light of social and technological changes.[119] So how viable is 'information command and control' in light of shifts in technology and culture?

'Command and control' of information: an uncertain future

As previous emergencies highlight, there are recurring problems achieving the objective of fast, clear, accurate information. The decentralised, fluid nature of information means that the ideal model of a clear, authoritative voice is difficult to sustain in the reality of an emergency.

The implications of failure to provide fast, consistent and accurate information in an emergency were starkly illustrated in the London bombings of 7 July 2005 (7/7). Although the 7/7 Review Committee Report found that the emergency response had been excellent in many respects, there were recurring flaws with communications between responders and to the wider public. For example, newspapers obtained information about the London Underground explosions within minutes of the first bombing.[120] However, official confirmation of the explosions was not provided until some two hours after the first bomb.[121] Official information initially confirmed (incorrectly) that there had been a 'power surge' on the Tube,[122] and was quickly overtaken by media coverage. As a result, news editors felt that the credibility of official information came into question.[123] The Committee found that the gap between known information and police confirmations can result in a loss of credibility and appear secretive.[124] Furthermore, the Committee found that basic information and advice on 7/7 should have been provided to the public sooner, within one hour.[125] The public information provided in the aftermath lacked practical instructions for people to follow, and some messages were contradictory.[126] The 'go in, stay in, tune in' message was disseminated for longer than necessary and this caused public confusion.[127] Subsequent incidents have also highlighted problems with timely dissemination of public information,[128] and with clarity of message.[129]

The provision of fast, clear and accurate information is not readily achievable in emergency circumstances. Establishing an accurate account of factual events on the ground is often intensely difficult in the confusion and uncertainty of a rapidly developing emergency situation.[130] Furthermore, the ideal may be frustrated by conflicts between standards of accuracy and speed.[131] Finally, information will be produced and disseminated by other bodies, particularly the media, which potentially undercuts and/or competes with the single, authoritative voice of responders.[132] As Twigg claims, 'in an age where people have access to more and more sources of information – in the media and in the internet – [the] controlling and centralising of information supply is no longer feasible'.[133] It is to these alternative sources of information that discussion now turns.

The media as an alternative source of information

CCA 2004 guidance provides detailed coverage of media handling as part of the public warning duty.[134] Because the media largely shapes the public's perception of an emergency,[135] it will also influence the emergency response.[136] *Emergency Preparedness* therefore indicates that media management can be categorised as 'other action in connection with an emergency' and that emergency plans should cover such matters.[137]

As well as the public service 'conduit' function during an emergency, the various forms of media will also undertake their usual reportage and, where necessary, the 'watchdog' function of holding officials to account.[138] Thus they are both 'vital allies' and 'potential adversaries'.[139] Guidance recognises that the media's emergency reportage will be concerned with the performance of the responders; 'the handling of the emergency, as well as the emergency itself, will all be part of the story'.[140] It advises Category 1 responders to plan ahead to deal with potential criticism and to avoid 'inconsistencies in presentation or message'.[141] The media's reportage role also frequently involves relaying the experiences of individuals involved in or affected by the emergency. By doing so the media may provide a platform for alternative narratives of the incident.[142] These may compete with the responders' singular message, particularly if there are varying perceptions regarding the efficacy of the response.[143]

Guidance indicates a clear concern to retain a degree of influence over the media coverage of the emergency, in order to ensure that the official narrative dominates. It draws a distinction between official and unofficial or informal information: 'It is important to be aware that information can be generated by official or unofficial sources'.[144] Failure to provide official information to the media will lead them to informal or alternative sources.[145] 'This may lead to a loss of messaging control ... and put [responders] in a position of having to defend themselves against unfounded criticism or inaccurate analysis'.[146] The assumption that unofficial information will be inaccurate and/or critical is present elsewhere in guidance.[147] As well as effective management of the media agenda, guidance also notes the importance of retaining ownership (and perceived ownership) of official information.[148] Yet, despite the preoccupation with such matters, media reportage may actually provide valuable information that potentially assists the emergency response. For example, the media helped to highlight the 'highly inadequate' response in crisis leadership during Hurricane Katrina.[149] Additionally, according to guidance, the media's real-time reportage, which will include individual experiences and/or analysis of the emergency response, is 'an important source of information' that will be fed back to Gold Command and presumably may be used as a basis for further action.[150]

In short, responders need the media (as a conduit to the public *and* as a source of intelligence) but simultaneously compete with it (in reportage mode) where we see a struggle for narrative control and informational authority. The clear, authoritative voice of responders becomes potentially just one among many.

The impact of new technologies and media

The emergence of digital technologies, particularly mobile phones and web-based media such as social networking sites (SNS) like Facebook and Twitter raises further challenges (as well as opportunities) regarding informing the public in emergencies. Government guidance does recognise the potential uses of social media in emergency communications with the public. Although it outlines certain risks with the medium,[151] it suggests that LRFs might consider producing social media protocols.[152] The volume of SNS-related guidance has gradually increased,[153] arguably reflecting mounting official recognition of the social significance of this and related technologies such as mobile alerting.[154]

As demonstrated in the 2004 Asian tsunami, new technologies enable the global dissemination of ground-level information in real time. *Emergency Preparedness* states:

> Mobile phones with cameras and other similar devices mean that the public are able to publish their own content. Within seconds of a disaster, pictures can be broadcast around the world, sometimes before the emergency services have had a chance to respond.[155]

This user-generated content (UGC) provides material for traditional media organisations, while also competing with them,[156] adding yet more 'voices' to the plethora, and ultimately increasing pressure on responders to provide timely information.[157] Yet this aspect of new technology is also particularly valuable to responders because UGC provides potentially vital independent intelligence during an emergency. Rather than linear, bottom-down communication, social media sites provide live two-way communications; they enable information to be communicated to the public,[158] but also information *from* members of the public to be gathered.[159] Yet they also crucially facilitate communications between members of the public.[160] Guidance states that 'Analysis of social media can ... aid situational awareness and better understanding of the working context',[161] and indeed SNS have been utilised by Australian responders to assist flood response.[162] For this reason guidance states that social media does potentially play a role in emergency preparation and response.[163] Furthermore, it is 'an area that emergency responders cannot ignore; whilst having little control over it'.[164] Consistent with this, recent research confirms that SNS adjust the traditional dynamic between 'commander' and the 'commanded' public:

> [They] present a disruptive technological challenge to the assumption of publics as passive recipients and consumers of one directional emergency broadcasts through traditional medial channels (e.g. radio and television).[165]

Despite limitations,[166] such technology is empowering and 'the citizen ... is recast from passive recipient to potential searcher, creator or collator of communication'.[167]

These developments exacerbate the difficulty responders face in dominating the wider coverage of an emergency, and aptly highlight the great disparity between the command model and the decentralised, fluid means by which information operates. The disparity was recognised by Exercise Watermark which found:

> The speed of social media means that some command and control structures are not flexible enough or fast enough to respond to public questions and this can undermine the authority of responders and the public's trust.[168]

These challenges are not restricted to informing the public in emergencies and indeed arise in other areas of law that involve ineffective attempts to control information, particularly in the digital domain, such as privacy injunctions,[169] and enforcing copyright to prevent illegal downloads.[170]

All these developments indicate that, as Twigg states, 'command and control of information is [now] unrealistic. The public are increasingly consumers of information from different sources, choosing what information to use and where to obtain it.'[171] As a result, Twigg argues that emergency planners will have to move away from command-and-control communicative approaches to alternatives that recognise these developments.[172] In short, 'public information and planning needs to become interactive, dynamic and responsive',[173] or even reflexive.[174]

Conclusion

Although clear, decisive leadership is an important aspect of an effective emergency response (alongside cooperation and flexibility), a command-and-control approach to information management in emergencies is increasingly outmoded and unviable, particularly in light of the potentially democratising forces of new technology. Information dissemination in emergencies does not follow a basic top-down, linear model; instead the flow of information is more complex and reciprocal, fluid and evasive. Government guidance does on occasion acknowledge this, such as by claiming that the monitoring of media and SNS can yield valuable intelligence for responders. Thus it becomes apparent that the authorities provide information to the media and public, the media provides information to the public and authorities and, finally, the public provides information to the media and authorities. This circulating information is used by all parties as a basis for action, therefore producing more information, which is used in turn as a basis for further (or modified) action.

Elsewhere, however, government guidance reveals the ongoing tendency to seek to cling onto an informational hegemony. For example, despite accepting that the media and SNS may provide useful information, elsewhere guidance indicates an underlying assumption that such information will be false or critical and that the best way to minimise this risk is to tactically feed information and manage media (old and new) so that the official narrative dominates. This is arguably justified at the height of an emergency, where responders might need people to take specific actions. But, beyond this specific situation, government needs to grasp that the 'official' narrative of events is not the only, or indeed the correct, one. Sometimes the authorities get it wrong, and sometimes official information about a crisis is not the only truth.

Government guidance on informing the public reveals an ongoing preoccupation with maintaining institutional authority (local or national) as much as empowering the public. Unlike in other forums,[175] 'official' information is not always *automatically* revered as definitive by the public and, arguably, nor should it be. Instead, informational authority must be earned through its accuracy and utility. The concern to provide authoritative information is at least partially linked to the (actual or perceived) authority that this fosters,[176] although such authority is ostensibly required to assist an effective response. So, ultimately, although government guidance advises engagement with these new technologies and recognises the fluid realities of the 'Information Society', its implicit aim is still too often command and control of information of sorts, albeit by more sophisticated, complex means.

Notes

1. These are identified as three of the four 'top-tier' threats to UK. HM Government, *A Strong Britain in an Age of Uncertainty: The National Security Strategy*, Cm 7953 (London, 2010), 27.
2. See, for example, Ben Wisner, Jean C. Gaillard, and Ilan Kelman, eds., *The Routledge Handbook of Hazards and Disaster Risk Reduction* (Abingdon: Routledge, 2012).
3. The Pitt Review, *Learning Lessons from the 2007 Floods*, 18.1, http://webarchive.nationalarchives.gov.uk/20100807034701/http:/archive.cabinetoffice.gov.uk/pittreview/thepittreview/final_report.html, (accessed November 10, 2013).
4. Civil Contingencies Act 2004 (Contingency Planning) Regulations 2005, SI 2011/615 (hereafter CCA(CP)R).
5. Listed in Civil Contingencies Act 2004 (CCA), Part I, Schedule 1.
6. Listed in Ibid., Part III, Schedule This should be Part 3, Schedule 1. (it is in Schedule 1, not Part III of the Act).
7. CCA(CP)R, reg 4.
8. Ibid., reg 4(1)–(3), reg 4(4)(3) and regs 4(4)(b), 4(7), 4(9), respectively. A list is available at https://www.gov.uk/local-resilience-forums-contact-details (accessed December 13, 2013).
9. Cabinet Office, *Emergency Preparedness* (London, 2012), 4.41–4.42, Box 4.4, https://www.gov.uk/government/publications/emergency-preparedness. See, further, Cabinet Office, *The Role of Local Resilience Forums: A Reference Document* (London, 2013).
10. Cabinet Office, *Emergency Preparedness*, 4.36–4.37.
11. CCA, s 2(1)(a); and CCA(CP)R, reg 13.
12. CCA(CP)R, reg 15. For example, the current CCR for Northumbria Local Resilience Area can be accessed at http://www.northumberland.gov.uk/pdf/CRR%207.1%20-%2023%20Apr%202012.pdf.
13. CCA, s 2(1)(d). For an example of an emergency plan, see the flood response plan for the Northumbria area, http://www.northumberland.gov.uk/idoc.ashx?docid=9ea03264-2f6d-4eca-9a48-d27b97e8163f&version=-1.
14. Cabinet Office, *Emergency Response and Recovery* (London, 2010), 4.2.19–33, https://www.gov.uk/government/publications/emergency-response-and-recovery; and Cabinet Office, *Responding to Emergencies: The UK Central Government Response* (London, 2013), 5.3–5.8, https://www.gov.uk/government/uploads/system/uploads/attachment_data/file/192425/CONOPs_incl_revised_chapter_24_Apr-13.pdf.
15. Cabinet Office, *Emergency Response and Recovery*, 4.2.23.
16. Particularly Cabinet Office, *Emergency Preparedness*; and Cabinet Office *Emergency Response and Recovery*.
17. CCA, ss 9(1)–(4).
18. CCA(CP)R, reg 26.
19. See https://www.gov.uk/government/collections/national-risk-register-of-civil-emergencies.
20. Cabinet Office, *Responding to Emergencies*, Section 2.
21. Ibid., Section 3.
22. Jim Broderick and Clive Walker, 'Applying "Civil Protection": A Review of the Civil Contingencies Regime in the UK' (paper delivered at the Society Guardian conference on 'Public Sector Emergency Planning and Business Continuity: Debating the Impact of the Civil Contingencies Act to Ensure your Community is better Prepared for the Future', London, May 21, 2007), 10, 11. See also Clive Walker and James Broderick, *The Civil Contingencies Act 2004: Risk, Resilience and the Law in the United Kingdom* (Oxford: Oxford University Press, 2006), 17.
23. These are the two most influential models that recur across government guidance and literature. For further discussion, see Timothy Sellnow and Matthew Seeger, *Theorizing Crisis Communications* (Oxford: Wiley–Blackwell, 2013), chap. 5, 114–19.

24. See, for example, *Burmah Oil v. Lord Advocate* [1965] AC 75(HL): 100 (Lord Reid); and Peter Hennessy, *The Prime Minister, the Office and its Holders Since 1945* (London: Penguin, 2001), 103.
25. *The Buncefield Incident 11 December 2005: The Final Report of the Major Incident Investigation Board*, Vol. 1, [153] http://www.buncefieldinvestigation.gov.uk/reports/ (accessed November 10, 2013).
26. Joint Committee on the Draft Civil Contingencies Bill, *Draft Civil Contingencies Bill* (2002–3 HL 184, HC 1074), Annex 7, 108–9.
27. Twigg, 'Disaster Risk Reduction, Mitigation and Preparedness in Development and Emergency Programming', 118, http://www.odihpn.org/hpn-resources/good-practice-reviews/disaster-risk-reduction-mitigation-and-preparedness-in-aid-programming (accessed November 10, 2013).
28. CCA(CP)R, regs 5, 7.
29. Broderick and Walker, 'Applying "Civil Protection"', 12; and Pitt Review, ES.96.
30. *Draft Civil Contingencies Bill*, [101]; Walker and Broderick, *The Civil Contingencies Act 2004*, 243; and Clive Walker, 'Governance of Critical National Infrastructure', *Public Law* (2008): 323.
31. Cabinet Office, *Communicating Risk* (January 2011), 52, https://www.gov.uk/government/publications/communicating-risk-guidance.
32. Thomas Hobbes, *Leviathan* (1651; repr., London: Penguin, 1985).
33. Ibid., 183–8.
34. Ibid., 192.
35. Ibid., 227.
36. Pierre Schlag, 'The Empty Circles of Liberal Justification', *Michigan Law Rev.* 96 (1997): 1, 26–7; and Jacques Derrida, *The Beast and the Sovereign*, Vol. 1 (Chicago: University of Chicago Press, 2009), 39–40. For a discussion of fear in the context of disasters, see Frank Furedi, 'New Dimensions: The Growth of a Market in Fear', in *Handbook of Disaster Research*, eds. Havidan Rodriguez, Enrico Quarantelli, and Russell Dynes (New York: Springer, 2006).
37. A term repeatedly used in HM Government, *A Strong Britain in an Age of Uncertainty.*
38. Ulrich Beck, 'Politics of Risk Society' and Anthony Giddens, 'Risk Society: The Context of British Politics', both in *The Politics of Risk Society*, ed. Jane Franklin (Cambridge: Polity, 1998), 12, 28.
39. Sellnow and Seeger, *Theorizing Crisis Communications*, 117–119.
40. CCA(CP)R, reg 4(1)–(3) (emphasis added).
41. Ibid., regs 44A, 47–50.
42. Broderick and Walker, 'Applying "Civil Protection"', 3 (emphasis added).
43. Scott Lash, *Critique of Information* (London: Sage 2002), 1.
44. John Perry Barlow, 'The Economy of Ideas: Selling Wine without Bottles on the Global Net', https://homes.eff.org/~barlow/EconomyOfIdeas.html (accessed October 10, 2013).
45. Ibid.
46. Manuel Castells, *The Rise of the Network Society* (Oxford: Wiley–Blackwell, 2010), chap. 6, 500; and Cabinet Office, *Emergency Response and Recovery*, 4.4.22, 8.3.3.
47. Zygmunt Bauman proposes 'fluidity' as 'the leading metaphor for the present stage of the modern era'. Bauman, *Liquid Modernity* (Cambridge: Polity, 2000), 2.
48. Lash, *Critique of Information*, 3, 18–21.
49. Ibid., 2, 18–20; and Castells, *The Rise of the Network Society*, xxxi–xxxii, chap. 7.
50. Lash, *Critique of Information*, 73–5.
51. Ibid., 4.
52. Ibid., 20, 26; and Castells, *The Rise of the Network Society*, xviii–xxvii, 501.
53. Lash, *Critique of Information*, 20.
54. Ibid., 39.
55. Ibid., 40, 41, 42.
56. Ibid., 35, xi, 25.
57. Brand, quoted by Barlow, 'The Economy of Ideas'.
58. Cabinet Office, *Emergency Preparedness*, 7.31.
59. Cabinet Office, *Emergency Response and Recovery*, 8.1.1 (emphasis added). See also 7.4.2.
60. CCA, s 2.

61. Ibid., s 2(1)(g).
62. Cabinet Office, *Emergency Preparedness*, 7.1, 7.6–7.7, 7.38–7.40; and Cabinet Office, *Emergency Response and Recovery*, 8.3.1.
63. CCA, s 2(1)(g); and CCA(CP)R, reg 20.
64. CCA(CP)R, reg 28.
65. Cabinet Office, *Emergency Preparedness*, 7.35, 7.26.
66. Ibid., 7.105.
67. Ibid., 7.72, 7.75.
68. Cabinet Office, *Emergency Preparedness*, 7.6.
69. Cabinet Office, 'Expectations and Indicators of Good Practice Set for Category 1 and 2 Responders' (April 2009), 27, https://www.gov.uk/government/publications/expectations-and-indicators-of-good-practice-set-for-category-1-and-2-responders.
70. CCA(CP)R, reg 32.
71. HM Government, *Preparing for Emergencies: What you Need to Know* (London, 2004), http://www.direct.gov.uk/prod_consum_dg/groups/dg_digitalassets/@dg/@en/documents/digitalasset/dg_176618.pdf.
72. CCA(CP)R, reg 29.
73. Ibid., reg 33.
74. Ibid., reg 34.
75. Cabinet Office, *Emergency Preparedness*, 7.22 Annex 7D step 3, 7.99.
76. Cabinet Office, 'Working with the Media', chap. 8 in *Emergency Response*. See also Clive Walker, 'The Police and the Mass Media in Emergencies', *Human Rights Review* 1, no. 1 (2011): 15.
77. Cabinet Office, *Emergency Response*, 8.3.5.
78. Ibid., 8.5.8–8.5.11.
79. Ibid., 8.5.3–8.5.5.
80. Ibid., 4.2.23, 4.2.25.
81. Ibid., 8.51–8.52.
82. National Steering Committee on Warning and Informing the Public, https://www.gov.uk/government/policy-advisory-groups/national-steering-committee-on-warning-informing-the-public (accessed November 10, 2013).
83. Cabinet Office, *Emergency Response*, chap. 13.
84. Ibid., 8.2, 13.4.1. See also Cabinet Office, *Emergency Preparedness*, Annex 7A; and Cabinet Office, *Responding to Emergencies*, 3.40–3.42.
85. Cabinet office, *Emergency Response*, 13.1.4.
86. See, for example, Cabinet Office, *Communicating Risk*, 3.4; and Cabinet Office, *Social and Behavioural Science Guidance for Local Resilience Forums in Planning and Conducting Civil Alerts* (London, 2012), https://www.gov.uk/government/publications/public-emergency-alerts-reviews-and-guidance.
87. CCA(CP)R, reg 27, reg 30 (emphasis added).
88. *National Risk Register* (see note 19), 1.16.
89. Cabinet Office, *Emergency Response and Recovery*, 2.6.6; and Cabinet Office, 'Expectations and Indicators of Good Practice', 56.
90. Thomas Drabek and David McEntire, 'Emergent Phenomena and the Sociology of Disaster: Lessons, Trends and Opportunities from the Research Literature', *Disaster Prevention and Management* 12, no. 12 (2003): 97, 98–9; and Wisner *et al*, *The Routledge Handbook*, 477.
91. London Assembly, *Report of 7 July Review Committee* (London, 2006), 4.1, recommendation 22, http://www.london.gov.uk/sites/default/files/archives/assembly-reports-7july-report.pdf.
92. Cabinet Office, *Emergency Preparedness*, 7.109.
93. Ibid., 7.103.
94. Use of images is one of six guiding principles of a communications strategy advocated in Cabinet Office, *Communicating Risk*, 46.
95. Pitt Review, ES.75
96. Ibid., 23.10–23.11, recommendation 68.
97. Cabinet Office, *Emergency Response and Recovery*, 2.6.7 (emphasis added). See also Cabinet Office, *Communicating Risk*, 47.
98. Guidance indicates that maintaining public credibility is an ongoing challenge for governments. Cabinet Office, *Communicating Risk*, 8, 52.

99. Cabinet Office, *Emergency Response and Recovery*, 8.1.1 (emphasis added). See also Cabinet Office, *Emergency Preparedness*, 7.4.2; and Cabinet Office, *Responding to Emergencies*, 3.40.
100. Cabinet Office, *Emergency Preparedness*, 7.42–7.43.
101. Ibid., 7.58. See also Cabinet Office, *Communicating Risk*, 46.
102. Cabinet Office, *Emergency Preparedness*, 7.59.
103. Ibid., 7.15, 7.60, 7.62, 7.86. See also Cabinet Office, *Communicating Risk*, section 6.4.
104. Ibid., 2.28, 7.16.
105. Cabinet Office, *Emergency Response*, 95 (case study).
106. Juan Carlos Villagran de Leon, 'Early Warning Principles and Systems', in Wisner *et al*, *The Routledge Handbook*, 486.
107. Cabinet Office, *Communicating Risk*, 18.
108. Cabinet Office, *Emergency Preparedness*, 7.57.
109. Cabinet Office, *Communicating Risk*, 8.
110. Cabinet Office, *Emergency Response*, 8.3.2. See also Cabinet Office, *Emergency Preparedness*, 7.32.
111. The BBC initiative, 'Connecting in a Crisis', http://www.bbc.co.uk/news/uk-15250977, (accessed December 11, 2013) seeks to preserve the communication channels of local and national radio and BBC TV news.
112. London Assembly, *Report of 7July Review Committee*, 6.2–6.3.
113. Pitt Review, 23.1, 23.2.
114. Cabinet Office, *Emergency Preparedness*, 7.121, but see 7.111, which states that responders cannot rely solely on the media to reach those involved: 'targeting will need to be more precise'. So, although it is effective, it has limitations.
115. London Assembly, *Report of 7July Review Committee*, 6.1, 6.4.
116. Cabinet Office, *Emergency Preparedness*, 15.3–15.4, 15.13.
117. Ibid., 7.125, 7.137. See also Ford Burkhart, *Media Emergency Warnings and Citizen Response* (Boulder, CO: Westview, 1991), 27, 31.
118. Twigg, 'Disaster Risk Reduction', 314.
119. John Sorensen and Barbara Vogt Sorensen, 'Community Processes: Warning and Evacuation', in Rodriguez *et al*, *Handbook of Disaster Research*, 198.
120. London Assembly, *Report of 7July Review Committee*, 6.18
121. Ibid., 6.20, 8.1.
122. Ibid., 6.19.
123. Ibid., 6.17.
124. Ibid., 6.21. See also Cabinet Office, *Communicating Risk*, 12.
125. London Assembly, *Report of 7 July Review Committee*, 6.11 onwards, recommendations 35, 41.
126. Ibid., 7.7, 8.6. Guidance acknowledges that inconsistent, contradictory messages can damage public confidence and be hard to repair. Cabinet Office, *Emergency Preparedness*, 7.62; and Cabinet Office, *Emergency Response*, 8.3.4.
127. London Assembly, *Report of 7 July Review Committee*, 8.1, 8.3, 8.17, recommendation 43.
128. Department for Environment, Food and Rural Affairs (DEFRA), *Exercise Watermark Final Report* (London, 2011), 3.111-3.112, recommendation 30, https://www.gov.uk/government/publications/exercise-watermark-final-report.
129. The Pitt Review found that public information during the 2007 floods was difficult to find, inconsistent (ES.99, 20.10), not explained clearly (ES.63) or too technical (10.24). In essence, 'there was no single authoritative voice' for warning and informing the public' (ES.62, recommendation 35).
130. Cabinet Office, *Emergency Response*, 8.3.3, 2.5.1, 4.4.17.
131. Cabinet Office, *Emergency Preparedness*, 7.83. See also London assembly, *Report of 7 July Review Committee*, 6.21.
132. Sellnow and Seeger, *Theorizing Crisis Communications*, 30.
133. Twigg, 'Disaster Risk Reduction', 177.
134. Cabinet Office, *Emergency Response*, chap. 8.
135. Cabinet Office, *Emergency Preparedness*, 7.118. See also Cabinet Office, *Communicating Risk*, 19.
136. Cabinet Office, *Emergency Preparedness*, 7.117.
137. Ibid., 5.12–5.13.

138. Pitt Review, 23.3; and Cabinet Office, *Emergency Preparedness*, 7.122. See also Burkhart, *Media Emergency Warnings*, 12.
139. Cabinet Office, *Communicating Risk*, 48.
140. Cabinet Office, *Emergency Preparedness*, 7.124.
141. Ibid., 7.123.
142. 'The multiple voices of disaster are easily lost over time, leaving posterity with only the official account of what took place. Such narratives need to be handled with care as they inevitably obscure how some use such events for their own profit.' Greg Bankoff, 'Historical Concepts of Disasters and Risks', in Wisner *et al*, *The Routledge Handbook*, 42.
143. Robert Littlefield and Andrea Quenette, 'Crisis Leadership and Hurricane Katrina: The Portrayal of Authority by the Media in Natural Disasters', *Journal of Applied Communication Research* 35, no. 1 (2007): 26, 42.
144. Cabinet Office, *Emergency Response*, 8.1.2. See also Cabinet Office, *Emergency Preparedness*, 7.122.
145. Cabinet Office, *Communicating Risk*, 56, 61; and Cabinet Office, *Emergency Response*, 8.5.6, 8.4.3.
146. Cabinet Office, *Emergency Preparedness*, 7.124.
147. Ibid., 7.122.
148. Cabinet Office, *Emergency Response*, 8.3.2. See also Cabinet Office, *Communicating Risk*, 60.
149. Littlefield and Quenette, 'Crisis Leadership and Hurricane Katrina'.
150. Cabinet Office, *Emergency Response and Recovery*, 8.5.1–8.5.2. Guidance states that media monitoring is one of six guiding principles of a communications strategy. Cabinet Office, *Communicating Risk*, 46.
151. Cabinet Office, *Emergency Preparedness*, 7.132.
152. Ibid. 7.133, 7.110.
153. This has been a particular focus for the National Steering Committee on Warning and Informing the Public (see note 82).
154. Cabinet Office, *Public Emergency Alerts: Mobile Alerting Trials* (London, 2013), https://www.gov.uk/government/publications/mobile-alerting-trials-for-public-emergencies/public-emergency-alerts-mobile-alerting-trials.
155. Cabinet Office, *Emergency Preparedness*, 7.128. For actual examples, see Cabinet Office, *Emergency Response and Recovery*, 95.
156. 'City Evacuations: Preparedness, Warning, Action and Recovery: Final Report of the DFUSE Project' (2013), 6, 5, http://www.cityevacuations.org/uploads/6/8/1/7/6817950/finalpublic.pdf.
157. Cabinet Office, *Emergency Response and Recovery*, 8.4.3.
158. Although traditional media are arguably a more effective initial warning mechanism. 'City Evacuations', 13–15, 37.
159. DEFRA, *Exercise Watermark*, 3.117–119; Cabinet Office, *Emergency Response and Recovery*, 8.3.14; and Cabinet Office, *Emergency Preparedness*, 7.131.
160. Castells terms this 'mass self-communication'. Castells *The Rise of the Networked Society*, xxx–xxxi. See also Sellnow and Seeger, *Theorizing Crisis Communications*, 127–131.
161. Cabinet Office, *Emergency Preparedness*, 7.131.
162. DEFRA, *Exercise Watermark*, 3.117.
163. Cabinet Office, *Emergency Preparedness*, 7.131. See also DEFRA, *Exercise Watermark*, vii and recommendation 32.
164. Cabinet Office, *Emergency Preparedness*, 7.130 (emphasis added).
165. 'City Evacuations', 6, 12. See also Sellnow and Seeger, *Theorizing Crisis Communications*, 73.
166. For example, reliance on social media to disseminate emergency information may marginalise groups without access, and also presupposes the resilience of this technology in such circumstances.
167. 'City Evacuations', 7.
168. DEFRA, *Exercise Watermark*, 3.123. See also 'City Evacuations', 5, 25.
169. For example, *CTB v. News Group Newspapers Ltd.* [2011] EWHC 1334 (QB).
170. Barlow, 'The Economy of Ideas'.
171. Twigg, 'Disaster Risk Reduction', 314. See also Sorensen and Sorensen, 'Community Processes', 186.
172. Twigg, 'Disaster Risk Reduction', 177.

173. 'City Evacuations', 25.
174. Ibid.
175. Rebecca Moosavian, 'Judges and High Prerogative: The Enduring Influence of Expertise and Legal Purity', *Public Law* (2012): 724, 740–748.
176. Cabinet Office, *Communicating Risk*, 46.

Emergency powers and the withering of the Royal Prerogative

Andrew Blick

Institute of Contemporary British History, King's College London, UK

A feature common to most governments in different time periods and places is that they possess exceptional powers applicable to emergencies. These authorities are necessary but vulnerable to abuse. Democracies seek to create and restrain such powers through constitutional texts and supplementary legislation. In the UK, lacking a written constitution, many of these executive entitlements have traditionally derived from the Royal Prerogative. This constitutional source comprises a group of residual monarchical authorities that have largely devolved in practice to ministers. As an historical entity the prerogative can only fully be understood in its more recent manifestations through taking a long-term perspective. It has often attracted controversy on the grounds that it is not subject to sufficient levels of what would now be termed accountability. At various points reform has been demanded and sometimes achieved. Since the later decades of the 20th century there has been an erosion of the prerogative, including its emergency powers facets. Parts of it have transferred to a statutory basis, while those elements that remain have become subject to enhanced legal and political constraints. This trend was part of a wider process of constitutional transformation traceable to the 1950s. Early in the twenty-first century the Royal Prerogative, already undergoing change, moved further to the centre of the constitutional reform agenda. This heightened salience arose from altered global security concerns, and associated developments in UK external policy. The Iraq conflict especially encouraged a focus on the powers under which the UK participated in this action. Calls for change, aimed at giving Parliament an enhanced role in war-making and over other matters, followed. Eventually a broad government reform package appeared. A convention regarding parliamentary consultation over engagement in armed combat developed. But features of the prerogative have persisted, including those allowing for government to act in the most perilous of circumstances. A thorough eradication of the Royal Prerogative could require a more complete overhaul of UK governmental arrangements, embodying them in a written constitution.

Emergency powers and the withering of the Royal Prerogative

Some functions of government are universal. They are so important to the basic cohesion of a society that they are common across different time periods, systems and geographic locations. One such group of responsibilities involves providing swift and coordinated responses to pressing threats. It extends also into preparing for such dangers and seeking to avert them. Individuals and agencies operating in this area often possess greater than usual powers and freedoms of action. This pattern reflects practical necessity. Extraordinary dangers and high stakes imply a need for, and justify the use of, more than normal countermeasures. Yet this logic is vulnerable to abuse, or at least the perception of unreasonable exploitation. Those in possession of authority may behave in a fashion more arbitrary than circumstances strictly require.[1] Indeed, the use or exploitation of extraordinary authorities can contribute to the destruction of the very political system they are intended to protect.[2]

Democracies often attempt to manage this dilemma through provisions in their constitutional texts, potentially supplemented by further enactments. Such mechanisms seek to provide the executive with the ability to anticipate and respond to exceptional circumstances, but at the same time to impose limitations upon it when doing so.[3] The UK is different. Famously it has no written constitution, in the sense of a single authoritative and comprehensive codification of governmental institutions, powers and relationships. If the UK ever obtains this kind of text, it is likely to contain provision for emergency powers. Indeed, during the 1650 s, the Commonwealth had two successive written constitutions (applying to England, Ireland, Scotland and Wales) that dealt with, among other matters, certain emergency authorities.[4]

Aside from the mid-17th century experiment, many emergency powers in English and UK constitutional history traditionally flowed from the constitutional source known as the Royal Prerogative. Consequently the issues of the legal basis for official policy towards emergencies and the prerogative are entwined. This article considers emergency powers in the widest sense, taking in the abilities to prevent, prepare for and respond to incidents domestically and abroad (including, for instance, through treaty-making and hostile overseas military deployments) and their existence under the Royal Prerogative. This constitutional legal source is an historic entity which can only fully be understood in its more recent embodiment using a long-term perspective. The paper discusses some of the controversy that has surrounded this particular constitutional source in the past, and historic calls for its reform. It then considers the erosion of the prerogative, especially in its emergency powers manifestations, that has taken place in recent decades. There is a particular focus on changes occurring during the twenty-first century. The article considers the causes of this trend (especially the shifting global security environment) and its implications.

The Royal Prerogative: definition, controversy and reform

Albert Venn Dicey described the Royal Prerogative as 'the residue of discretionary or arbitrary authority, which at any given time is legally left in the hands of the Crown'.[5] In history the monarch exercised the prerogative personally. Today it has largely been delegated in practice to ministers. Definition of this non-statutory, common law power is doubly uncertain: doubt surrounds both the power in general and particular aspects of it. First, in seeking a broad description, one interpretation might be that it is all the powers delineated by the common law that are possessed by the executive (or 'Crown', the equivalent though not identical term employed in UK constitutional language). Another version is that the Royal Prerogative comprises the authorities and immunities, again identified by the common law, that are different from those in the possession of an ordinary person. Often commentators hold these versions to be competing, although it may be that they do not contradict each other.[6]

Further confusion arises because there is no definitive or legally binding list of prerogative authorities, and a modern comprehensive account does not exist even in the most weighty constitutional law texts.[7] Their existence or otherwise can only be settled when arising for adjudication by a court – not a frequent occurrence. In 2009 the Ministry of Justice produced the fullest official picture to date, based on the first internal government attempt at an itinerary of the prerogative toolkit. It discerned four general categories.[8]

First was 'Ministerial prerogative powers'. A number of functions in this group had potential or direct relevance to emergencies. They included the conduct of diplomacy; treaty-making and treaty ratification; the acquisition and ceding of territory; providing, denying and revoking passports; and giving diplomatic protection to people from the UK who were in other countries. The prerogative provides for the making of declarations of war and for participation in conflicts not involving declarations of war (no such declaration has been issued since 1942, against Siam, although the UK has participated in many conflicts subsequently). When war is over, the UK can make peace under the same authority. Responsibility for the armed forces domestically and abroad also derives from the prerogative (though it is intertwined with statute). So too does the ability to direct the armed forces in helping law enforcement agencies, or other public bodies, perhaps during industrial action. The prerogative allows, the Ministry of Justice contended, for regulating trade with enemies; confiscating the property of neutrals in wartime ('angary'); and claiming vessels and property of enemies acquired at sea ('Prize'). The review noted that there were other special powers applicable in times of 'grave national emergency', giving the example of being able to interfere with private property. It is also possible for the executive to initiate public inquiries under the prerogative, including into its handling of emergencies.

A second category labelled 'Constitutional/personal prerogatives' incorporated the hiring and firing of ministers; appointing the prime minister; dismissal of a government; providing Royal Assent to legislation; and conferring honours. Falling under this heading is another emergency right: to 'require the personal services of subjects in case of imminent danger'.

A third group was 'Powers exercised by the Attorney General'; and a fourth 'Archaic prerogative powers'. The continued existence of rights in this latter group, the government observed, was doubtful. Within this category of uncertain status were further emergency powers: to impose service in the Royal Navy; to prevent people, in extremity, from leaving the territory; and in wartime to detain, remove or impose restrictions on an 'enemy alien'. Finally came 'Legal Prerogatives of the Crown'.[9]

In all, then, the Royal Prerogative provides the executive with an important range of discretionary powers of potential or direct relevance to emergencies. Providing for action through the Royal Prerogative rather than via a means such as a written constitution or statute law has certain arguable consequences. It perhaps means there is less clarity about these powers than otherwise would be the case. Moreover, Parliament or some other font of democratic legitimacy has by definition not authorised or scrutinised the authorities concerned. The formal exclusion of the legislature is crucial, given that since the 17th century it has indisputably been the 'sovereign' body within UK constitutional understandings. Yet Parliament is neither the source for, nor has it had a specific formal role in approving the use of, the prerogative. However, another tenet in the UK is that the executive is, or can be, accountable to Parliament for anything that it does (with limited exceptions); and parliamentary authorisation is needed for the expenditure that any policy requires, even if carried out under the prerogative.[10] The judiciary has the ability to determine whether a prerogative power exists. But there has been a tradition for the courts not to scrutinise the way in which the prerogative is exercised. In the words of James I (James VI of

Scotland) the manner in which monarchs deployed the 'absolute Prerogative of the Crown' was 'no Subject for the tongue of a Lawyer'.[11] This principle has, however, changed in recent years, as shall be explained. Arguably the judiciary would be even more likely to become involved in such areas were they set out and limited in a more precise text. However, a degree of reluctance on the part of courts to rule on the particulars of executive emergency action is likely to remain under any system.

Disputes over the Royal Prerogative – including its emergency component – are a recurring theme of English and UK constitutional history. The great struggles of the 17th century involved contesting views of what was the appropriate balance between the Royal Prerogative, on the one hand, and the authority of the King in Parliament, on the other hand. The ability of the monarch to take emergency action, both domestically and externally, was an important point of contention. Support for an enhanced parliamentary role in armed combat sometimes comes from those opposed to a particular conflict or to war in general. Yet during this period those who challenged the prerogative were often advocates of aggressive military support for the Protestant cause in Europe.[12]

Events of the 17th century served to curb more extreme applications of the Royal Prerogative and helped establish the principle that it was subordinate to Acts of Parliament. Under the terms of the Bill of Rights 1689, while the prerogative continued to exist, it could only be retrenched and not expanded. However, notwithstanding this general tenet, it has not always been clear whether particular prerogative powers survive. An Act must abolish them either explicitly or by necessary implication. When the courts rule on the persistence or otherwise of an authority, they sometimes possess a degree of discretion in determining whether it can coexist with a statute that does not expressly revoke it and whether the prerogative power under consideration needs to be reinterpreted in changed circumstances. In the process, the judiciary might, in the view of some, perform a creative role.

During the course of the 18th and 19th centuries the Royal Prerogative came increasingly in practice to be exercised by ministers. While, as noted above, in theory ministers were generally accountable to Parliament, including for the exercise of the Royal Prerogative, there were problems with the realisation of this concept. Parliament did not have the specific engagement in the prerogative that it did in the domestic law-making process. In 1872, in the introduction to the second edition of his seminal work, *The English Constitution*, which first appeared in 1867, Walter Bagehot commented on this position. He noted that treaties could be as significant as legislation but, while the approval of parliamentarians was necessary 'to every word of the law', their assent was not needed 'even as to the essence of the treaty'. This disparity, Bagehot held, was '*prima facie*, ludicrous'.[13] On 19 March 1886 the House of Commons debated a motion that it was 'not just or expedient to embark in war, contract engagements involving grave responsibilities for the nation, and add territories to the Empire without the knowledge and consent of Parliament'. This proposition fell, but only by a margin of four votes.[14] The reference in this resolution to the 'knowledge' of Parliament is revealing. The legislature not only lacked specific powers with respect to the prerogative. Parliamentarians might not even know that a conflict had broken out before it was too late to exert influence; and the executive was under no obligation to inform it of all its treaty commitments, or of other actions carried out under the Royal Prerogative.

The perception that secret international undertakings had helped precipitate the European, and then world, conflict that began in 1914 was an important influence on the objectives of the Union of Democratic Control (UDC), formed at the outset of the First World War. This cross-party pressure group campaigned for foreign policy to be more open, and subject to greater parliamentary oversight. A founder member of UDC, Arthur

Ponsonby, took particular interest in the parliamentary dimension of its programme. He developed a scheme that included a requirement for explicit parliamentary authorisation for the detail and generality of any kind of treaty or international agreement. The approval of the legislature would also be necessary for declarations of war. Ponsonby advocated the establishment in Parliament of a Foreign Affairs Committee to scrutinise policy. Ponsonby later became Parliamentary Under-Secretary of State for Foreign Affairs in the first Labour government of 1924. From this position he was able to give a limited expression to the UDC agenda. On 1 April 1924 he undertook to the House that the government would seek to lay every treaty, once signed, before both Houses of Parliament for 21 days before ratification took place. If an opposition party requested a debate on the agreement, the government would find the time for it in the parliamentary schedule.

Eventually this commitment developed into a convention known as the 'Ponsonby Rule'. While it proved generally successful in ensuring that the texts of treaties were made available to Parliament, it was not effective as a means of facilitating plenary debates, let alone votes, on their substance. In practice the legislature becomes involved in international agreements in this way when and if they require changes to domestic legislation. The UK tends towards the 'dualist' rather than 'monist' constitutional model, meaning that international agreements to which it is party are not automatically directly integrated into the UK legal hierarchy, something that requires parliamentary approval. Legislators therefore do have the opportunity to discuss and vote on bills giving expression to international commitments. But this practice is not the same as considering the agreements themselves, and only takes place if such legislation is required by the nature of the treaty. It is, however, also possible for an Act of Parliament to oblige parliamentary approval in specific areas, as is the case with the European Union Act 2011.

Accelerating change

As the comments of Bagehot, and the demands of parliamentarians and the UDC show, important portions of the prerogative, including those with an external affairs and security dimension, persisted well past the 17th century. But over time Acts of Parliament began to appear on this territory. For instance, important security measures in the two World Wars were implemented on a statutory basis, albeit with a high degree of delegated ministerial order-making authority.[15] In recent decades the Royal Prerogative in general, and the facets of it relating to emergencies and defence in particular, have undergone sustained pressure and incursion, arising from a potent combination of tendencies. Sometimes limitations upon the prerogative have come about without loudly being proclaimed as such. Yet latterly commentators and even policy-makers have tended to depict this outcome as a virtue in itself. This process has occurred in an altered environment affecting upon many features of the UK constitution, and leading to substantial change. The disastrous Suez venture of 1956, and the decline in the UK as a global force of which it was emblematic, helped inaugurate a trend of pronounced domestic criticism of arrangements for the governance of the country. Official institutions, furthermore, were more likely to challenge each other. These trends were coupled with a wider decline in social deference. In such a climate the Royal Prerogative appeared increasingly anachronistic, and lacking in both sufficiently precise definition and democratic legitimacy. The transformed position of the UK in the world drove constitutional change in other ways. Participation in the accelerating process of continental integration, including the Council of Europe and what is now known as the European Union, had profound implications.

There has been further dramatic change in the field of intelligence and security.[16] The Interception of Communications Act 1985 provided for activities including phone-tapping, previously seemingly carried out under the prerogative. As well as creating a new basis for these powers, the Act established safeguards against their abuse. In 1989 the Security Service Act gave some statutory existence to the Security Service (commonly known as MI5), which had since its formation in 1909 been a creature of the prerogative. The Secret Intelligence Service (popularly labelled MI6) had origins in the same year. It acquired statutory sanction through the Intelligence Services Act 1994, which further provided for the Government Communications Headquarters (GCHQ). The 1989 and 1994 Acts both created oversight bodies for the agencies concerned, including commissioners and tribunals. The 1994 Act established a committee of parliamentarians, the Intelligence and Security Committee, appointed by the prime minister after consulting with the leader of the opposition, to monitor the three agencies. It became a full committee of Parliament, with an expanded brief, under the Justice and Security Act 2013, Part I. The pattern set by this legislative activity was for a power to transfer away from the prerogative and at the same time become subject to greater constraints.

Changes beyond the immediate realm of statute also took place. The intelligence and security agencies were at the centre of events once again. In a landmark case in 1984 the Law Lords considered the official policy of a ban on trades union membership at GCHQ.[17] In the process they asserted the appropriateness of the courts scrutinising the manner of the exercise of the prerogative, though finding that in this particular instance they should defer to the executive. The Law Lords thereby expounded a significant new principle, supplanting the earlier version of the role of lawyers as expressed by James I/VI.

The conduct of executive government in general, even though parts of this set of tasks continue to be matters of prerogative, is dealt with increasingly in published official statements of principles and rules. They include the *Ministerial Code*, first made publicly available in 1992 (after previously existing as an internal document) under the earlier title of *Questions of Procedure for Ministers*; and the *Civil Service Code*, issued in 1996.[18] The executive drafts and owns these documents. They lack statutory provenance and independent enforcement mechanisms. But they might conceivably be used against the government in judicial review proceedings in future; they already serve at times politically to constrain its freedom of action under the prerogative. For instance, premiers may feel obliged to remove ministers who seem flagrantly to have violated the *Ministerial Code*.

Though not a parliamentary committee, the formation of the Intelligence and Security Committee, under the Intelligence Services Act 1994, section 10, was connected to developments in Parliament. From 1979 a new group of select committees sat in the House of Commons, normally shadowing individual departments. Providing more systematic select committee coverage of the executive than existed before, the activities they inquired into included those engaged in under the prerogative. Among their number was a Foreign Affairs Committee, representing in a sense the fruition of another part of the Ponsonby/UDC programme of more than half a century previously. This committee system has continued to develop. In 2000, in an adaptation of the Ponsonby Rule, the government made undertakings to give select committees an enhanced role in the scrutiny of treaties. The Liaison Committee, which comprises the chairs of all these bodies, has since 2002 taken evidence from the prime minister twice a year, sometimes quizzing him on the use of the prerogative for emergency purposes. The Backbench Business Committee, established in 2010, provides for debates in parliamentary time not controlled by the government, and that facility can be used for consideration of matters of Royal Prerogative. As part of the same reform package that included the introduction of this Committee, arrangements commenced for the members and chairs of Commons select committees to be elected by MPs rather than determined by the whips, strengthening their independence from the government and from partisan orientation. More broadly MPs are displaying a growing propensity to vote against party instructions, including – as will be shown – over the use of prerogative emergency powers.

The Royal Prerogative in the twenty-first century

The Labour government that took office in 1997 embarked on a vast constitutional reform programme. Areas of change included devolution for parts of the UK, the Human Rights Act 1998 and the Freedom of Information Act 2000. Yet, initially, although Labour had shown some interest in it earlier in the 1990s,[19] the government did not for a decade contemplate serious curtailment of the Royal Prerogative.

Beyond the government, this issue was receiving some attention. In 1996 the Liberal Democrat Peer Lord Lester had introduced a bill to provide for parliamentary oversight of treaties. He withdrew it after obtaining a guarantee that under the Ponsonby Rule treaties would in future be accompanied by an explanatory memorandum, to aid parliamentary understanding of their implications. Professor Rodney Brazier, writing in 1999, argued that the principle of ministers being responsible to Parliament did not provide a satisfactory limitation on their potential misuse of the prerogative, and called for a series of reforms.[20] Shifting dynamics and concepts in global politics, and the UK approach to them, added impetus to such proposals, particularly in the area of armed conflict. Even before the terrorist attacks on the US of 11 September 2001, the UK had, since the end of the Cold War, been participating more frequently in hostile military action, for instance in the Gulf in 1991 and against atrocities in Kosovo in 1999. By the beginning of the twenty-first century there were signs of the development of a new doctrine of international intervention, echoing ideas from earlier centuries, that came to be labelled 'Responsibility to Protect'.[21] The UK was in the vanguard of efforts to shift moral and legal conceptions in this direction. But what were the domestic constitutional implications? A more regular and wider occurrence of military action, coupled with a different philosophy regarding its use – which might in itself reinforce the tendency of rising instances of armed combat for the UK – could suggest a need for new powers and mechanisms for the legislature in overseeing such operations. However, a concerted effort to bring about change had not yet occurred.

Emergent security imperatives, then, served to accelerate the withering of the Royal Prerogative. Rising concern about international terrorism of the brand associated with the 9/11 attacks was an important – though not sole – motive for the Civil Contingencies Act 2004.[22] Discussed elsewhere in this special issue, it overhauled emergency powers provision, that previously derived both from legislation now deemed outmoded and the Royal Prerogative, to take into account threats including climate change and fuel protests, as well as terrorism. Part II of the Act sought to ensure that government was equipped to respond to incidents threatening 'serious damage' to 'human welfare', the 'environment' or 'security' in the UK. It made specific reference to 'war, or terrorism' (section 19). It provided for senior ministers, in an emergency as defined under Section 19 of the Act and in specific conditions set out in Section 21, to make 'emergency regulations' (section 20). Section 22 defines these regulations as directed towards 'preventing, controlling or mitigating' the emergency, and as being able to make 'provision of any kind' that could be achieved through an Act of Parliament or the Royal Prerogative. The precise implications of this Act for the Royal Prerogative are discussed later.

Also related to the events of 9/11 was an increasingly belligerent US approach to global policy that found its most controversial manifestation in the invasion of Iraq of 2003. In the UK participation in this action, a subject of great political contention, drew attention to the power under which government committed the armed forces to combat. The Labour MP Graham Allen, a leading figure among the parliamentary opponents of the war, was one who specifically linked this campaign to the issue of the prerogative. In September 2002, before the operation commenced, he called for the introduction of a requirement for parliamentary approval for entry into military conflict, whenever possible in advance of action.[23] Various similar proposals appeared. An important influence on these schemes was the US War Powers Resolution (or War Powers Act) of 1973, which seeks to provide Congressional oversight of military deployments not involving actual declarations of war.[24]

Partly thanks to the efforts of Allen, the Royal Prerogative moved to the centre of the constitutional reform agenda. The House of Commons Public Administration Select Committee (PASC), chaired by another Labour MP, Tony Wright, issued a report on the subject of the Royal Prerogative in March 2004.[25] It accepted that special executive authorities were needed, particularly for responding to emergencies, but argued that they required more effective oversight by Parliament than existed at the time. The Committee called on the government to commence a public consultation on the prerogative powers, containing 'proposals for legislation' that would bring about 'greater parliamentary control over' the entirety of these authorities as wielded by ministers. The consultation should, PASC argued, also include explicit plans for bringing about 'full parliamentary scrutiny' of decisions to enter into armed combat, of the concluding and ratifying of treaties, and the issuing and revoking of passports.

The government was not receptive to the PASC recommendations, emphasising in its response the need to preserve flexibility in areas where urgent action might be required.[26] But pressure for action over the prerogative, and especially over the war powers, continued to mount. In 2006 the House of Lords Select Committee on the Constitution concluded that the Royal Prerogative was no longer an appropriate basis on which to 'deploy armed force overseas'.[27] But rather than advocating specific abolition of the prerogative in this area, it supported its regulation through a 'parliamentary convention'.[28] Both the Commons and the Lords would have a role, although if they disagreed the view of the Commons would take precedence over that of the Lords. Under the arrangement envisaged by the Committee, the government would obtain consent before engaging in 'actual or potential armed conflict'.[29] It would set out what were its 'objectives', how long the engagement was likely to last, the scale of the commitment, and the legal case for the operation. In special circumstances the government could seek retrospective support, within seven days or as early as possible. It would keep Parliament updated about developments, and if the parameters changed significantly, seek a fresh mandate.

Then on 15 May 2007, with the end of the Blair premiership looming, the Commons passed a resolution proposed by HM Opposition about the votes held by the government on the Iraq war in 2002 and 2003.[30] It held that in 'seeking and obtaining the approval' of the Commons for the operation, the government had set a precedent the reversal of which in future was 'inconceivable'. It called for a 'more explicit' position for Parliament, taking into account the need for discretion in some cases, and for the government to consult and produce 'detailed proposals'.

While the Blair government was reluctant to act in the areas highlighted by PASC and others, it had introduced – as discussed above – one major piece of legislation, the Civil Contingencies Act 2004, with implications for the Royal Prerogative. A further area of progress – though tortuously slow – was towards a Civil Service Act. In theory the Labour

Party was committed to shifting the power for the management of Whitehall from a prerogative to a statutory grounding. While this policy was not a priority, in 2004 the government issued a draft bill for consultation.[31] In the area of public inquiries the Inquiries Act 2005 created a new, more comprehensive, statutory framework for this activity, but left prerogative reserve powers intact. The Armed Forces Act 2006 introduced a new statutory basis for Royal Navy inquiries. Further development, over the Civil Service and other parts of the prerogative, had to await the transition to a new prime minister.

Gordon Brown and the *Governance of Britain*

The arrival of Gordon Brown as prime minister opened possibilities for retrenchment of the Royal Prerogative on a number of fronts. Brown intended to make an impact at the outset of his premiership by unveiling a programme of constitutional reform. In July 2007 the Ministry of Justice issued a paper, *The Governance of Britain*, setting out the main features of the agenda.[32] It made various proposals designed to secure a more accountable executive, to achieve democratic renewal and to reinvigorate national identity. Prominent within these plans were reforms to the Royal Prerogative. The paper noted that the executive had for hundreds of years 'been able to exercise authority in the name of the Monarch without the people and their elected representatives in their Parliament being consulted'.[33] It concluded that such an arrangement was 'no longer appropriate in a modern democracy',[34] in which the executive should derive its authority from the public, via Parliament. The government described how it wanted to create a requirement for parliamentary approval for the entry of the armed forces into combat, though allowing for 'swift action to protect our national security' and not compromising 'operational security or effectiveness'.[35] The government also expressed the intention to place the Ponsonby Rule on a statutory footing. It promised a Civil Service Act, setting out the main principles of this institution. The paper proposed a review to consider whether the Royal Prerogative as a whole should be 'codified or brought under statutory control'.[36] However, it specifically ruled out changes to the status of the 'legal prerogatives of the Crown or the Monarch's constitutional or personal prerogatives',[37] although it might change the way in which ministers decided how to advise the monarch to deploy these powers.

Various more detailed official papers followed. In October 2007 the government consulted on how to give Parliament a role in war-making – including whether such arrangements should be provided for in convention or by statute – and on enshrining the Ponsonby Rule in an Act of Parliament.[38] Then in March 2008 a draft *Constitutional Renewal Bill* appeared which included in its clauses provision for treaties and the Civil Service.[39] It did not propose statutory regulation of parliamentary consultation over armed conflict, but accompanying the draft bill was the text of a detailed draft Commons resolution on this subject. It excluded the idea of having to seek retrospective approval for deployments too urgent or sensitive to consult on in advance; Parliament would simply be informed about what had taken place. Under the proposed scheme the Commons would be able to vote on an intervention, while the Lords could only debate it.

The most important outcome of the Royal Prerogative dimension of the *Governance of Britain* policy initiative came with the Constitutional Reform and Governance Act 2010. Among other provisions, Part I of the Act gave the power to manage the Civil Service, previously derived from the prerogative, a statutory foundation. Part II contained stipulations about the ratification of treaties, part of the purpose of which was legally to enact the Ponsonby Rule.[40] The provisions did not abolish the prerogative in this area, but made its exercise subject to statutory regulation. They stated that ratification of a treaty (with some exceptions) could not take place unless the text had been published and laid before Parliament for 21 sitting days. During this period, if the Commons or Lords passed a resolution against ratification, a minister – if wishing to proceed – would have to produce an account of why it was necessary to do so. Such a statement would be sufficient to counter an objection by the Lords. But if the Commons had voted against the agreement, a further 21-day period would then begin, during which the Commons (but not the Lords) could block ratification once again. The minister would then be able to respond again, and so on (section 20). A minister could, however, bypass the process altogether if of the opinion that to do so was required for exceptional reasons, but not after the Commons or Lords had already resolved against a treaty (section 22).

While the 2010 Act may have encapsulated the Ponsonby Rule, it did not provide for the earlier more radical Ponsonby proposal that treaties should come into force only if they received the express and detailed approval of Parliament. The presumption was that silence amounted to assent. Moreover, the Act did not deal with how a debate and vote might take place. There was no change to parliamentary procedure or structures accompanying the Act specifically to enable the effective exercise of this new power. An important feature of the Ponsonby/UDC programme was the establishment of parliamentary mechanisms that would provide a meaningful role in overseeing government external policy. While the reform of the select committee system of 1979 and its subsequent development represented some movement in this direction,[41] in relation to treaties there remained much scope for further development. The proposal sometimes made for a specific parliamentary treaties committee had yet to come about.

War powers: further developments

The war powers component of the *Governance of Britain* agenda went no further. Nonetheless, a convention that Parliament (specifically the Commons) has some kind of general right to be consulted expressly on military action, preferably in advance, seems to be developing.[42] New impetus has come from the geopolitical consequences of the 'Arab Spring', a regional surge of political upheaval that gained force early in 2011. On 29 August 2013 the government brought about an emergency recall of the Commons from recess to discuss and hold a vote on a substantive motion setting out the case for an intervention in Syria over the use of chemical weapons.[43] Prime Minister David Cameron emphasised that, if MPs supported this resolution, the government would still need to wait until United Nations weapons inspectors had reported and the UN had deliberated before the UK could participate in an action. Even then, he would seek a further Commons vote. This approach meant that the government was seeking explicit approval from the Commons early in the decision-taking process and would involve it again, presumably in the same way, further on. Of equal importance, the Commons then showed that the procedure was more than a ritual. It withheld its support, and in so doing demonstrated that the power it possessed in this area was real. The Commons effectively blocked the UK from participating in any US initiative to take military action in Syria, regardless of the collective view of the UN Security Council. In theory the prime minister could have pressed on anyway, since he was under no legal obligation to abide by the will of the Commons in this area. But he judged such a course of action politically impractical: 'It is very clear tonight that, while the House has not passed a motion, the British Parliament, reflecting the views of the British people, does not want to see British military action. I get that, and the Government will act accordingly.'[44]

Retrospectively this constitutional development began in 2002 when the government was contemplating taking part in an invasion of Iraq. At this point, although Parliament could become engaged in matters of armed conflict, it did not possess the rights of prior approval it then began to acquire. The circumstances of the Iraq operation helped instigate change for two reasons. First, as already noted, the contentious nature of this action helped trigger demands for greater levels of legislative control over war-making. Second, and ironically, the very controversy surrounding the prospect of intervention in Iraq seems to have influenced the government to subject the mission to an exceptionally high level of parliamentary consultation, in advance of its full commencement. Inside the government the then Foreign Secretary, Jack Straw, and the Leader of the House of Commons, Robin Cook, pressed for the involvement of the Commons.[45] Substantive debates and votes took place in the Commons on 25 November 2002, 26 February 2003 and 18 March 2003.[46] The latter two occasions saw the largest rebellions by members of a governing party in modern parliamentary history.

The embedding of the convention has not been smooth. In 2011 the UK took part in military operations in Libya. A debate and vote on a substantive motion took place in the Commons, but on 21 March, after the action was announced on 18 March.[47] Only eight days beforehand, on 10 March, the Leader of the House of Commons, Sir George Young, had told the House that a convention had emerged that the Commons should be able to discuss deployments before they occurred (except in special circumstances), and that this principle would apply to Libya.[48] The justification the prime minister offered for seeking only retrospective approval was that the government needed to proceed with urgency once the United Security Resolution was obtained.[49] He perhaps honoured the idea of a convention in accepting a need to explain why it did not apply on this occasion. Moreover, possibly reflecting the vulnerability of the government on this count, the Foreign Secretary, William Hague, committed in the debate of 21 March to introducing legislation that would create an obligation for seeking parliamentary approval.[50] This pledge represented a step beyond previous Conservative policy that a convention was sufficient in this area.

Yet, although a statutory option had apparently come into play, the convention concept continued to gain ground. The *Cabinet Manual*, published in October 2011, is a statement by the executive of a variety of constitutional and governmental rules and practices. It referred to the government acceptance of 10 March 2011 of the existence of a convention.[51] In January 2013 the government once again negatively recognised the rule by insisting that deployments in Mali of that month, on which no vote was held, would not involve participation in hostilities and therefore did not engage the emerging rule.[52] Positive entrenchment came later in the year, when the government sought – and failed to obtain – approval in principle for action in Syria.[53]

Conclusion: the remnants of the prerogative

Events surrounding war and emergency powers show that some of the momentum for reform of the Royal Prerogative that grew markedly under the Labour administration of Gordon Brown survived the 2010 transition to the Coalition government. Indeed, the set of authorities regarded as more personal to the monarch has been clipped, a possibility excluded under Labour. Previously dissolutions of Parliament, triggering general elections, were granted by the head of state. Negotiations leading to the establishment of the Conservative/Liberal Democrat coalition produced an agreement to legislate for parliaments with a standard length of five years. This deal came into force via the Fixed-term Parliaments Act 2011.[54] Section 3 of the Act stated that Parliament could not be dissolved other than in accordance with the Act itself. The prerogative power to grant dissolutions was at an end, though the monarch still had a role under the Act in setting the dates for early elections and summoning Parliament after elections.

While the Royal Prerogative is diminished in its scope and legitimacy, it still endures. The full review promised as part of the *Governance of Britain* led to the most comprehensive official account of these powers yet published, described near the beginning of this paper. In the 2009 report containing this list, the government stated that, beyond its existing specific plans, it did not intend at this point to pursue the more comprehensive reform it had floated in 2007. Executive powers over passports continue to derive from the Royal Prerogative, subject to non-statutory rules. The government continues to be able to initiate public inquiries under the prerogative, a power it has deployed heavily in relation to the Iraq war, by setting up the Hutton, Butler and Chilcot inquiries.[55] This approach could be seen as enabling the executive to some extent to skew the way in which it is investigated to suit its purposes, and as rivalling parliamentary inquiries. While a convention exists that Parliament has certain rights over armed conflict, this constitutional understanding is not directly legally enforceable; in many ways it is loose and potentially subject to varying interpretations. A government might claim, as it did over Libya in 2011, that the urgency of acting precludes advance consultation. It is not clear how the convention might apply to cyber-attacks or the use overseas of Unmanned Aerial Vehicles (UAVs, also known as 'drones') or how to distinguish a support or training role for troops from a full-blown foreign engagement. As a consequence, the executive may still enjoy great flexibility. The legislation that William Hague promised in 2011 has proved slow to materialise. Important resistance to such a measure, in the Gordon Brown period and subsequently, has come from inside military circles, within which there is a wider apprehension about growing legal encroachment upon the conduct of war. A particular concern among those who share this perspective involves the use of the Human Rights Act to challenge violations of the European Convention on Human Rights.[56]

In July 2013 the House of Lords Select Committee on the Constitution concluded that the convention that had emerged on parliamentary involvement over military action was sufficient reform. The Committee held that arrangements should not become more formal either through a Commons resolution or an Act. It argued that to go beyond the existing approach to consultation with the legislature could raise difficulties in defining what qualified as a hostile engagement; might involve a loss of valuable flexibility in political and democratic processes; and could create a problematic exposure to legal action, including in relation to operational decisions. However, the Committee did recommend that a future edition of the *Cabinet Manual* include a detailed account of internal government procedures for the taking of decisions over armed conflict.[57] In its response to the Committee the government – agreeing it would at an appropriate point act upon the request about the *Cabinet Manual* – appeared to back away from the idea of legislation in this area.[58]

Diplomacy continues to be conducted under the prerogative. So too does treaty-making, although Parliament can in theory prevent the ratification of international agreements under the Constitutional Reform and Governance Act. The work of Graham McBain shows that a variety of prerogative powers with emergency connotations, though their deployment has long-since ceased, have not clearly been abolished.[59] They include the right to force subjects into the army or navy, to oblige people to provide accommodation and subsistence to members of the armed forces, to supply letters guaranteeing safe conduct, to order subjects to return to the country, to prevent them from leaving it, and even to enforce martial law.

Furthermore, in its 2009 review, the Ministry of Justice discussed the possible continued existence of 'Powers in the event of a grave national emergency including those to enter upon, take and destroy private property'. It considered whether the comprehensive coverage of the Civil Contingencies Act 2004, though not explicitly replacing the prerogative, had supplanted it. As noted above, section 22 (3) of the Act stipulated that emergency regulations were able to make any provision that would be possible through an Act of Parliament or the Royal Prerogative. Perhaps this description indicated that the latter source of power continues to be relevant. The Ministry of Justice concluded that there might be circumstances in which time constraints were too severe to allow adherence to the procedural requirements of the Act for the issuing of regulations. It could also be that a failure in communications could force personnel at some level to act without specific authorisation under the terms of the Act. Under these conditions it could be legal to draw upon the Royal Prerogative. The government concluded that this possibility should remain open.[60] One of the most basic functions of government, to act decisively when all else – including statutory arrangements for emergency regulations – had failed, continued to rest on the Royal Prerogative.

To conclude, emergency powers and the Royal Prerogative from which a number of them derive have long occupied a contested area within the UK/English constitution. In this territory different requirements for the public good overlap. Tension between them is inevitable. On the one hand, a degree of executive discretion is needed in some areas. On the other hand, official actions should be politically and legally accountable. Yet there is also mutual dependency between the two imperatives. Properly conceived emergency authority exists to protect rather than subvert a constitutional order, which must in turn to some extent accommodate exceptionality in the interests of its own security.

Generally the reform of the Royal Prerogative has been a vehicle for reducing the degree of executive discretion, making the emergency powers it wields subject to more constraints. External pressures, such as conflict in Europe in the 17th century, or more recently a changing global security agenda, may motivate reform, especially when interacting with domestic political tendencies that are also conducive to change. The existence of Royal Prerogative powers has become progressively more difficult to justify and sustain. Even where it survives it is subject to enhanced judicial and parliamentary oversight. The drive to change the emergency prerogatives should be seen as a component and important motivator of a wider process of reduction of this constitutional power source, for instance through a Civil Service Act and legislation for fixed-term parliaments. The Royal Prerogative has withered. However, as a consideration of the residual emergency powers demonstrates, it has not yet disappeared altogether. This final obliteration might involve the adoption of a full written constitution of the sort that provides the basis for such authorities in other democracies. Perhaps the same forces that have prompted recent drastic changes in our system of government point towards this outcome.

Notes

1. See, for example, A.W. Brian Simpson, *In the Highest Degree Odious: Detention without Trial in Wartime Britain* (Oxford: Clarendon, 1994). In this instance, the powers involved were derived from statute, not from the Royal Prerogative.
2. As took place in Germany in the 1930s. See Michael Burleigh, *The Third Reich: A New History* (Basingstoke: Pan, 2001).
3. See, for instance, arrangements under the 1978 Constitution of Spain. Victor Ferreres Comella, *The Constitution of Spain* (Oxford: Hart, 2013).

4. See Samuel Rawson Gardiner, ed., *The Constitutional Documents of the Puritan Revolution* (Oxford: Clarendon Press, 1906).
5. Albert Venn Dicey, *Introduction to the Study of the Law of the Constitution* (Indianapolis: Liberty Fund, 1982), 282.
6. See Maurice Sunkin and Sebastian Payne, eds., *The Nature of the Crown* (Oxford: Oxford University Press, 1999).
7. See, for example, Robert Blackburn, 'Monarchy and the Personal Prerogatives', *Public Law* 3 (2004): 546; and A.W. Bradley and K.D. Ewing, *Constitutional and Administrative Law*, 15th ed. (London: Longman, 2010), chap. 12.
8. Ministry of Justice, *Review of the Executive Royal Prerogative Powers: Final Report* (London: Ministry of Justice, 2009). I have omitted from this account of the review prerogative powers abolished since its publication, especially those relating to the management of the Civil Service.
9. Ibid., 30–4.
10. See Bill of Rights 1689, art 4; *The Case of the King's Prerogative in Saltpetre* (1606) 77 ER 1294; *R v. Hampden* (1637) 3 Cobb.St.Tr. 826; David Lindsay Keir, 'The Case of the Ship-Money', *Law Quarterly Review* 52 (1936): 546; Michael Mendle, 'The Ship Money Case, the Case of Shipmony, and the Development of Henry Parker's Parliamentary Absolutism, *The Historical Journal* 32 (1989): 513; and *Burmah Oil Co. Ltd. v. Lord Advocate* [1965] AC 75.
11. Interpretation and citation in Martin Loughlin, *Foundations of Public Law* (Oxford: Oxford University Press, 2010), 379.
12. See Jonathan Scott, *England's Troubles* (Cambridge: Cambridge University Press, 2000).
13. Walter Bagehot, *The English Constitution* (London: Fontana, 1963), 289. Arthur Ponsonby subsequently reproduced this remark in the work cited below.
14. Arthur Ponsonby, *Parliament and Foreign Policy* (London: Union of Democratic Control, 1915), 8.
15. See Defence of the Realm Act 1914; and Emergency Powers (Defence) Act 1939.
16. See Jon Moran and Mark Pythian, eds., *Intelligence, Security and Policing Post-9/11: The UK's Response to the 'War on Terror'* (London: Palgrave Macmillan, 2008); and S. McKay, *Covert Policing: Law and Practice* (Oxford: Oxford University Press, 2011).
17. *Council of Civil Service Unions v. Minister for the Civil Service* [1985] AC 374. See also Clive Walker, 'Review of the Prerogative: The Remaining Issues', *Public Law* (1987): 62; and Rebecca Moosavian, 'Judges and High Prerogative: The Enduring Influence of Expertise and Legal Purity', *Public Law* (2012): 724.
18. See now Cabinet Office, *Ministerial Code* (London: Cabinet Office, 2010), https://www.gov.uk/government/publications/ministerial-code.
19. Keith Syrett, 'Prerogative Powers: New Labour's Forgotten Constitutional Reform', *Denning Law Journal* 13 (1998): 111.
20. Rodney Brazier, 'Constitutional Reform and the Crown', in Sunkin and Payne, *The Nature of the Crown*, 337–62.
21. See UN Security Council Resolution 1674 (2006), para 4, which 'Reaffirms the provisions of paragraphs 138 and 139 of the 2005 World Summit Outcome Document regarding the responsibility to protect populations from genocide, war crimes, ethnic cleansing and crimes against humanity'; and C.G. Badescu, *Humanitarian Intervention and the Responsibility to Protect: Security and Human Rights* (London: Routledge, 2010).
22. See Clive Walker and James Broderick, *The Civil Contingencies Act 2004: Risk, Resilience and the Law in the United Kingdom* (Oxford: Oxford University Press, 2006).
23. Graham Allen, MP, 'A Degree of Democratic Deficit', *Daily Telegraph*, September 26, 2002, http://www.telegraph.co.uk/comment/letters/3582038/A-degree-of-democratic-deficit.html.
24. 50 U.S.C. ss 1541–1548. See John Yoo, *The Powers of War and Peace: The Constitution and Foreign Affairs after 9/11* (Chicago: University of Chicago Press, 2005); and Peter Irons, *War Powers: How the Imperial Presidency Hijacked the Constitution* (New York: Metropolitan Books, 2005).
25. House of Commons Public Administration Select Committee, *Taming the Prerogative: Strengthening Ministerial Accountability to Parliament* (2003–04 HC 422).
26. http://webarchive.nationalarchives.gov.uk/+/http://www.dca.gov.uk/pubs/reports/prerogative.htm, 2004 (accessed December 27, 2013).

27. House of Lords Select Committee on the Constitution, *Waging War: Parliament's Role and Responsibility* (2005–06 HL236-I), para 103.
28. Ibid., para 108.
29. Ibid., para 110.
30. *Hansard* (House of Commons), vol. 460 col. 582.
31. Cabinet Office, *A Draft Civil Service Bill: A Consultation Document*, Cm 6373 (London: Stationery Office, 2004). The degree of priority may be judged by the fact that the legislation was not finalised until April 2010 in the shape of the Constitutional Reform and Governance Act 2010 (see below).
32. Secretary of State for Justice and Lord Chancellor, *The Governance of Britain*, Cm 7170 (London: Stationery Office, 2007).
33. Ibid., para 14.
34. Ibid.
35. Ibid., para 26.
36. Ibid. para 50.
37. Ibid., para 24.
38. Ministry of Justice, Ministry of Defence and Foreign and Commonwealth Office, *War Powers and Treaties: Limiting Executive Powers*, Cm 7239 (London: Stationery Office, 2007).
39. Ministry of Justice, *The Governance of Britain – Constitutional Renewal*, Cm 7342 (London: Stationery Office, 2008), 3 vols.
40. See Jill Barrett, 'The United Kingdom and Parliamentary Scrutiny of Treaties: Recent Reforms', *International & Comparative Law Quarterly* 60 (2011): 225.
41. For consideration of a treaty prior to ratification by a select committee, see, for example, House of Commons Defence Committee, *UK/US Defence Trade Cooperation Treaty* (2007–08 HC107).
42. See Michael Head, 'Calling out the Troops and the Civil Contingencies Act: Some Questions of Concern', *Public Law* (2010): 340; and Nigel White, 'International Law, the United Kingdom and Decisions to Deploy Troops Overseas', *International & Comparative Law Quarterly* 59 (2010): 814.
43. *Hansard*, vol. 566 col. 1425.
44. Ibid., cols 1556–7.
45. See, for example, Jack Straw, *Last Man Standing* (London: Macmillan, 2012).
46. *Hansard*, vol. 395 col. 47; vol. 400 col. 265; vol. 401 col. 760.
47. Ibid., vol. 525 col. 700.
48. Ibid., vol. 524 col. 1066.
49. Ibid., vol. 525 col. 613.
50. Ibid., col. 799.
51. Cabinet Office, *The Cabinet Manual* (London: Cabinet Office, 2011), para 5.37.
52. *Hansard*, vol. 557 col. 1059–60.
53. See, further, Claire Mills, *Parliamentary Approval for Deploying the Armed Forces: An Update* (London: House of Commons Library, 2013).
54. See Mark Ryan, 'The Fixed-term Parliaments Act 2011', *Public Law* (2012): 213; and Raymond Youngs and Nicklaus Thomas-Symonds, 'The Problem of the "Lame Duck" Government: A Critique of the Fixed-term Parliaments Act', *Parliamentary Affairs* 66 (2013): 540.
55. See *Report of the Inquiry into the Circumstances Surrounding the Death of Dr David Kelly* (2003–04 HC 247); *Review of Intelligence on Weapons of Mass Destruction* (2003–04 HC 898); and http://www.iraqinquiry.org.uk.
56. For a flavour, see Tom Tugendhat and Laura Croft, *The Fog of Law: An Introduction to the Legal Erosion of British Fighting Power* (London: Policy Exchange, 2013).
57. House of Lords Select Committee on the Constitution, *Constitutional Arrangements for the Use of Armed Force* (2013–14 HL 46).
58. See also two reports by the House of Commons Political and Constitutional Reform Committee: *Parliament's Role in Conflict Decisions* (2010–12 HC 293); and *Parliament's Role in Conflict Decisions: An Update* (2013–14 HC 649).
59. Graham McBain, 'Abolishing Obsolete Crown Prerogatives Relating to the Military', *Nottingham Law Journal* 20 (2011): 20; and McBain, 'Abolishing Obsolete Crown Prerogatives

Relating to: Martial Law, Conscription and Billeting', *International Law Research*, 1, no.1 (2012): 13.

60. Ministry of Justice, *Review of the Executive Royal Prerogative Powers: Final Report* (London: Ministry of Justice, 2009), 30–4.

The governance of emergency arrangements

Clive Walker

School of Law, University of Leeds, UK

Part II of the Civil Contingencies Act 2004 ('emergency powers') has remained almost a dead-letter and has never been fully implemented. The first part of this paper will reflect upon the contents and performance of Part II. The second part will seek to explain the powerful official aversion to it. The reasons for this aversion relate not only to the political implications arising from the implied admission of a loss of control or failure of resilience but also to a disinclination to comply with the mechanisms for limitation and accountability in Part II. The latter point is evidenced by the production of draft Bills which set a lower threshold for use whenever a suitable type of emergency arises. Two such Bills have appeared: the Draft Detention of Terrorist Suspects (Temporary Extension) Bill 2011 and the draft Enhanced Terrorism Prevention and Investigations Measures Bill 2011. This growing trend will be analysed and criticised with respect to the potentially diminished governance of resilience and emergency measures.

Introduction

We can readily conjure for ourselves all manner of nightmares, actual or predicted. They appear to provide excellent reasons why we should divert money from education or health towards contingency planning and resilience. Unlike in movies such as *The Day After Tomorrow,*[1] one hopes that governments will 'not make a drama out of a crisis', as was once the advertised boast of the insurance company, Commercial Union.

In the face of calamity and disaster the Civil Contingencies Act 2004 (CCA 2004) has emerged as the principal legal response in the UK,[2] although a considerable amount of additional 'sectoral' laws have also emerged to handle specific crises. Notable specialist laws relate to: terrorism, such as occurred on 7 July 2005;[3] floods, which caused widespread damage and dislocation in 2007;[4] and animal and human diseases such as Foot and Mouth in 2001,[5] and H1N1 influenza in 2009.[6] Beyond these specialist examples, it is expected that the systematic approach of Part I of the CCA 2004, imbued with the concepts both of 'risk management' and networked governance, will avert crisis as well as mitigate any damage.

We know from experience that not all problems can be averted or confined to a minimum, although the dangers should not be exaggerated. These recent British calamities were not on the scale of the 9/11 attacks or of natural events such as Hurricane Katrina or Typhoon Haiy, but they are amplified by what sociologist Frank Furedi has called the 'culture of fear',[7] which leads into the politics of fear – the talking up of risks either because of diminished tolerance to risk or because of political manoeuvring to gain status as visionary and protective.

Even for large-scale risks, many of the state's responses will still fall mainly under Part I of the CCA 2004 and will thereby be localised. However, the focus of this paper is on Part II of the CCA 2004, which specifies a comprehensive scheme of emergency powers to be invoked in response to the most serious live emergencies. This paper will provide critical analysis in response to the following questions: what does Part II offer in time of emergency in order to assist with dealing with the crisis; how has Part II been elaborated since enactment in 2004; when might Part II have been invoked to deal with serious emergencies which have occurred since 2004 and why has it remained in abeyance; and what legal measures or devices have been invoked instead of Part II?

What does Part II of the CCA 2004 offer in time of emergency?

Part II of the CCA 2004 is the successor to the Emergency Powers Act 1920 (plus the Emergency Powers Act (Northern Ireland) 1926).[8] This skeleton legislation enabled Her Majesty in Council to proclaim that a state of emergency existed and thereupon to make regulations to deal with that emergency. The 1920 Act was used exclusively to deal with the impacts of major industrial strikes and was invoked on 12 occasions in Britain. But the last occasion was in 1979, and it was increasingly perceived as an inadequate response to contemporary conditions. Defects included its emphasis upon dangers to the supply or distribution of food, water, fuel, light or the means of locomotion, thereby ignoring contemporary risks such as animal and human health diseases, terrorism, or technological vulnerabilities. Next, there was perceived to be a need to be responsive within more defined geographical bounds rather than requiring the whole of Great Britain or Northern Ireland to be affected. There was also a need to take account of the arrangements for devolution in Scotland, Wales and, to a lesser extent, Northern Ireland. The final issue was a requirement for more specific respect for individual rights, in so far as required by the Human Rights Act 1998. There may have been some temptation simply to repeal the legislation – de-industrialisation and privatisation within the UK meant that its original purposes were no longer pressing. However, there was a sharp recent reminder of its potential utility during the severe dislocation caused by the widespread protests about fuel duties in 2000.[9]

As enacted, Part II embodies the opposite emphasis from Part I of the CCA 2004 (which is discussed further in this issue by Moosavian). Rather than localised planning, with the role of central government being unspoken, Part II emphasises central not local involvement, as well as action not planning. The substance of Part II (sections 19 to 31) involves four elements.[10]

The first element is a statement which enumerates the 'live' emergencies covered by Part II. 'Emergency' is defined by section 19(1) to include (a) events and situations which threaten serious damage to human welfare or (b) the environment, or (c) war or terrorism which threaten serious damage to security. By section 19(6), the event or situation mentioned in subsection (1) may occur inside or outside the UK. The differences from the Part I definition in section 1 are relatively minor, which, on the face of it, shows remarkable reticence, given that the two Parts serve quite distinct purposes. One would also expect the thresholds in Part II to be much more demanding, since its powers are potentially more intrusive into normal life and individual rights. One explicit difference concerns the scale of events which the two Parts address. In Part II the words, 'Part or region' are used instead of 'place' in Part I. A 'Part' means England, Northern Ireland, Scotland and Wales. A 'region' means one of nine regions. So the Part II powers cannot be triggered by localised problems, which might only affect a specific 'place' within a Part or region and which are the basis for planning and preparations under Part I.

The concern about the breadth of the definition leads to the second element. The scheme of the CCA 2004 is to maintain parity in the two sets of definitions in each Part but to insert into Part II more preconditions before use (section 21) and more limitations over use (sections 22, 23 and 26) – all reflecting the 'Triple Lock' strategy of demanding seriousness, necessity and geographical proportionality. Thus, the first condition (section 21(2)) is that an 'emergency' (as defined in section 19) has occurred, is occurring, or is about to occur. The second condition (section 21(3)) is that the regulations are 'necessary' to make provision for the purpose of preventing, controlling or mitigating an aspect or effect of the emergency. The third condition (section 21(4)) is that the need for regulations is urgent. As for limits, the minister making emergency regulations must have regard to the importance of ensuring that Parliament, the High Court and the Court of Session are able to conduct proceedings in connection with (a) the regulations, or (b) action taken under the regulations (in section 22(5)). By section 23 the regulator must be 'satisfied' (a) that the provision is appropriate for the purpose of preventing, controlling or mitigating an aspect or effect of the emergency in respect of which the regulations are made, and (b) that the effect of the provision is in due proportion to that aspect or effect of the emergency; in addition the regulator must bear in mind geographical restraint (subsection (2)). Section 23 imposes substantive restraints – regulations may not (a) require a person, or enable a person to be required, to provide military service, or (b) prohibit or enable the prohibition of participation in, or any activity in connection with, a strike or other industrial action. By section 23(4) regulations may not (a) create an offence other than one of the kind described in section 22(3)(i) (in other words, failing to comply with a provision of the regulations or directions or orders under a regulation or obstructing a person carrying out a function under the regulations), or (b) create an offence other than one which is triable only before a magistrates' court or, in Scotland, before a sheriff under summary procedure, or (c) create an offence which is punishable (i) with imprisonment for a period exceeding three months, or (ii) with a fine exceeding level 5 on the standard scale, or (d) alter procedure in relation to criminal proceedings. By section 23(5) emergency regulations may not amend Part II of the CCA or the Human Rights Act 1998. A further limit appears in section 26. Emergency regulations must lapse after 30 days, beginning with the date on which they are made, or at an earlier time specified in the regulations.

The third element – the bulk of Part II – is an almost boundless power in section 22(1) to issue emergency regulations. Emergency regulations may make any provision which the issuing authority is satisfied is appropriate for the purpose of preventing, controlling or mitigating an aspect or effect of the emergency in respect of which the regulations are made. In this way, there are long lists of the types of powers and provisions which may be required by emergency regulations, as well as the shorter list of what may not be included in section 23.

The final element relates to organisation. Most of the architecture remains unspecified by the legislation, although there is some enlightenment via the Cabinet Office guidance, *Emergency Response and Recovery*.[11] One exception is that, when Regional Recovery Teams are engaged in response mode, this may give rise to the Regional Civil Contingencies Committee (RCCC) to coordinate the regional response. If the emergency becomes exceptionally serious, then the RCCC would be chaired by the Regional Nominated Coordinator (RNC). If Part II of the CCA 2004 comes into operation, the RNC is recognised in section 24. Otherwise, the guidance suggests that, when an emergency arises, the response and recovery phases will principally be dealt with by local Strategic Coordinating Groups. The management and coordination of local operations involve the familiar Gold, Silver and Bronze tiers. Gold command for response, usually headed by the police, takes the form of a Strategic Coordinating Group (SCG) when there is an emergency with a substantial impact or lasting over a prolonged period.[12] In recovery phases, Gold Command may pass to the local authority, with the establishment of a Recovery Co-ordinating Working Group.[13] Other aspects of local organisation include the care of those affected. A local authority crisis/emergency centre may be formed, along with a Humanitarian Assistance Centre.[14]

As for central guidance, the Lead Government Department (LGD) is advised to maintain and equip an emergency control centre.[15] When an emergency occurs, the range of duties include: acting as the focal point for communications between central government and the local strategic coordinating group; producing a situation report; producing a handling plan which advises as to the capacity of the LGD; applying the LGD resources and coordinating the support from other departments and agencies; taking decisions and actions and providing assistance and advice; publishing information for the public and the media; and providing accountability to Parliament and any inquiry.[16] As well as planning and prevention, the Cabinet Office also contains a crisis coordination centre which can operate when the crisis reaches beyond the interests or capabilities of one government department. In this way the Cabinet Office Briefing Room (COBR) will be the location for meetings about the involvement of central government.[17] But the Cabinet Office guidance emphasises at the outset that LGD involvement is to be viewed as exceptional, where the scale or complexity of the crisis so demands,[18] and that most emergencies should be handled locally, since 'local responders ... remain the basic building block of the response to an emergency.'[19]

Part II of the CCA 2004 represents the most powerful and extensive peacetime legislation ever enacted. It reflects a determination to respond to extreme risk and to do so by methods which involve command and control rather than a broad governance approach. Indeed, it contains the tools for dismantling civil society. It looks like a security state's dream – it could even deliver some powers denied in battles over anti-terrorism legislation.

How has the CCA 2004 been elaborated since enactment?

Two major pieces of work have been undertaken to elaborate on Part II. The first is the Cabinet Office guidance document, *Emergency Response and Recovery*, now in its fifth edition (the first edition was issued in 2005 and the latest in 2013). The most important updates followed the commitment given by the Home Secretary in response to the Coroner's report on the 7 July London bombings.[20] It also reflects the demise of the system of regional government offices after 2010 (although some offices remain open for resilience purposes in Leeds, Birmingham, Bristol and London).[21] The document has grown to 233 pages – twice the size of the original issued in 2005, which was 104 pages. The guidance aims to provide greater clarity and to indicate good practice, including lessons learnt from experience. The objectives are said to be:

- a shared understanding of the multi-agency framework for emergency response and recovery at the local level, and the roles and responsibilities of individual organisations;
- a shared understanding of the role of local, regional and national levels in emergency response, and how they will work together; and

- a common frame of reference, especially concepts and language, for those involved in responding to emergencies.[22]

Despite its burgeoning page count, its limitations should be underlined. It remains far briefer than the equivalent guide on Part I (*Emergency Preparedness*).[23] The *Emergency Response and Recovery* guidance is essentially directed towards delineating emergency response and recovery at the local and regional level and the roles and responsibilities of individual organisations. The national role is also explained. But only chapter 14, the final chapter, covers emergency powers with reference to Part II. Part II is there described as 'a last resort', and there is even the warning that: 'There must be no expectation that the government will agree to use emergency powers and planning and response arrangements should assume that they will not be used.'[24]

The second major area of substantive work of potential relevance relates to military intervention.[25] The military are an alluring and important resource in an emergency as part of a coordinated response; regular forces may possess dedicated assets specifically related to emergencies, such as bomb disposal teams and search and rescue facilities.[26] Non-dedicated assets might comprise non-specialist manpower or bridging and pumping equipment. The working out of clear doctrines and structures in such emergencies is undertaken in the Ministry of Defence document, *Operations in the UK*.[27] The document emphasises that the military are 'a last resort'.[28] This statement underscores guidance on Part I of the CCA 2004 that military involvement must be exceptional:

> assistance is provided on an availability basis and the armed forces cannot make a commitment that guarantees assistance to meet specific emergencies. It is therefore essential that Category 1 and 2 responders do not base plans and organise exercises on the assumption of military assistance.[29]

The *JDP02* documentation mainly relates to the mechanisms for call-out, the types of potential work and also the allocation of costs. In administrative terms the involvement of the military will mainly fall under the generic heading of Military Assistance to the Civil Authorities (MACA) in the form of Military Aid in the Civil Community (MACC), which involves unarmed support in emergencies, whether under the CCA 2004 or under section 2 of the Emergency Powers Act 1964, which addresses the legality of military commands to engage in non-military work but is not accompanied by any special measures as to the military use of force or the exercise of special powers.[30]

As well as these guides as to processes, powers and practices, the growing emphasis upon contingency planning also led to the setting up of what was originally called a Reserve Reaction Force,[31] and is now labelled the Civil Contingency Reaction Forces (CCRF, of which there are 13 within the Territorial Army).[32] The core of a CCRF is formed around a Territorial Army infantry battalion (roughly 500 personnel). There is a provision for call-out to deal with an emergency in section 52 or 56 of the Reserve Forces Act 1996, subject to ministerial authorisation. These units can provide general duties support, which may be used to supplement the local civil emergency response capabilities, but

> They have no dedicated assets beyond the normal equipment held by their parent unit(s). If called to deploy, they would draw on the local divisional equipment pool for any additional equipment needed to suit the emergency requirement e.g. shelters, camp beds, tables, tarpaulins, sandbags, torches. They would be provided with the necessary resources to complete their task.[33]

The House of Commons Defence Committee has expressed concern that the 'Reserves rather than the regulars may now be the first port of call' in an emergency, especially as they will have limited training and their availability is not instant.[34] Availability has also been depleted by duties in Afghanistan and Iraq.[35] It should also be noted that the legislative framework was not augmented, for example to provide special powers or to deter obstruction. One might compare the Emergency Workers (Obstruction) Act 2006, which provides for offences of obstruction either of specified emergency workers (fire, ambulance, coastguard, or Royal National Lifeboat Institution) when responding to an emergency or those assisting them.[36] This interesting development seeks to facilitate generic emergency relief, but it is not comprehensive.[37] It adds to, rather than supplants, existing laws such as pre-existing offences relating to assaults or obstructions of enforcement officers.[38] Unlike some of those offences, it is confined to circumstances when these workers/officials are responding to an emergency and not to the protection of critical infrastructure *per se*.

Aside from the limitations in the military assets available for emergency work, the latter's use in what is termed 'United Kingdom Operations (UK Ops)' raises issues concerning the laws and other processes governing their involvement. There is an absence of statutory regulation of joint planning and processes for extraordinary scenarios for which the central state is inevitably involved, such as terrorist attack.[39] None of these matters is explicitly covered by, or related to, the CCA 2004, and there has been little interest in expanding the legislative framework.

The only spark of interest in expanding emergency laws has concerned the idea of consultation with Parliament on war powers prerogatives, which emerged in a paper from the Ministry of Justice, *The Governance of Britain*,[40] as part of Prime Minister Gordon Brown's initial enthusiasm for constitutional renewal. The final paper in this reform agenda, issued in 2009, *The Governance of Britain: Review of the Executive Royal Prerogative Powers: Final Report*, did actually point to shortcomings in Part II of the Civil Contingencies Act:

> The time needed to make emergency regulations under the 2004 Act could make compliance impractical in some emergency scenarios ... it is extremely unlikely that emergency regulations could be put in place quickly enough to be of any use ... There could be circumstances in which Government officials who were not able to contact a Minister to fulfil the requirements of the Civil Contingencies Act might need to take immediate action, in reliance on the prerogative, to combat an urgent threat.
>
> In practice, therefore, the Royal prerogative might need to be relied on in place of the Civil Contingencies Act in particularly extreme and urgent circumstances and on a strictly time-limited basis; indeed these may be the only ways in which it can lawfully be used.[41]

In short, 'These [prerogative] powers may provide a vital ability to act where there is insufficient time to put statutory provisions in place'.[42] The Brown administration concluded that they must be retained, and the Coalition Government has shown no inclination to resile from that stance.[43]

The previous government was similarly dismissive of the need for reform of its apparatus. For example, it rejected the criticism of the Home Affairs Committee that COBR was too reactive and its activation process was too uncertain.[44]

Another area where there was an opportunity missed to build on Part II of the CCA 2004 concerns the Critical National Infrastructure.[45] There is now a Cabinet Office Sector Resilience Plan for Critical National Infrastructure, covering priorities, plans, roles and responsibilities, as well as more specific plans for individual sectors.[46] But these documents are woefully brief and give no indication of any role for Part II.[47]

Perhaps the most significant indicator of all of the relative neglect of Part II is that it was placed outside the remit of the Civil Contingencies Enhancement Programme (CCAEP) because, according to the Cabinet Office:

> (i) Part 2 is untested in practice: emergency regulations under Part 2 have never been invoked since coming into force; (ii) recommendations from recent reviews focus on refinement of Part 1 – there is no specific evidence to suggest that changes to Part 2 of the CCA are required; and (iii) a separate workstream, being led by CCS [the Civil Contingencies Secretariat] is addressing operational aspects relating to invoking CCA emergency powers.[48]

The separate stream of work conducted by the CCS on operational aspects relating to invoking CCA 2004 emergency powers produced a revised version of the document, *Emergency Response and Recovery*, but with no proposed or sustained major alterations.[49]

The conclusion to this survey of the development of the CCA 2004 Part II is that, while regulation within government may be generally growing,[50] there remains in the planning and response to extreme situations regulation through what one commentator terms 'club government' which is informal, oligarchic and secretive.[51]

When might the CCA 2004 have been invoked and why was it not?

The CCA 2004 Part II came into force on 10 December 2004.[52] There may be two subsequent events where one might have expected some resort to Part II. The first was the response to the terrorist attacks on 7 July 2005. Admittedly terrorism is the risk above all which has been the subject of special sectoral legislation, public expenditure, planning and exercising. But one could imagine some role for emergency regulations related to people and traffic management, the requisition of property for casualties or equipment, and the processing of witnesses. However, the Home Office's reflection on the events, *Addressing Lessons from the Emergency Response to the 7 July 2005 London Bombings* (London, 2006) was that the Civil Contingencies Act was relevant primarily to the 'Prepare' leg of CONTEST – for building resilience under Part I.[53] There was no admission that Part II was required, although some emergency machinery, such as COBR, was very much in evidence. Thus, the CCA 2004 is viewed as relevant only to the 'Prepare' phases of counter-terrorism.[54]

The second event which largely dodged the Part II provisions concerned the great floods of 2007. The reports of the Pitt Review revealed a number of tensions in the development of networked arrangements dealing with dire emergencies.[55] For example, the floods of 2007 triggered the activation of Regional Civil Contingencies Committees (RCCC) in the South West region on the 23 and 24 July 2007. According to the Pitt Review, this was the first activation of an RCCC since the establishment of the CCA 2004.[56] However, the role and performance of the RCCC were criticised. The underlying reasons for the activation of the RCCC were not clearly understood on the part of response agencies. Perhaps, more seriously, confusion also arose over the authority of the RCCC and its relationship with central government. While conceding that the timing of the activation and disestablishment of the RCCC was 'sensible', the Review recommended that the Cabinet Office (and others) take a more proactive role in explaining the role of RCCCs to local responders.[57]

The Pitt Review also criticised the LGD role in the management of the floods. Its First Report noted that the LGD (in this instance, the Department for Environment, Food and Rural Affairs – DEFRA), 'took time to settle into an effective pattern' and found its position as LGD complicated by the split of responsibilities between itself and the Environment Agency.[58] The Review noted that the activation of the COBR mechanism was a result of a 'broadly accurate' estimation of the extent and severity of the flooding on the part of the Environment Agency and 'played an important role in the achievement of improved performance. Departments felt that the response was better coordinated and more focused than had been the case in June.'[59] Nevertheless, there appears here something of a confused picture, with the LGD taking some time to delimit its own role and then effectively to respond. One might question, therefore, whether the activation of COBR in July was as 'timely' as described, given its dependence on estimates emanating from an Environment Agency/DEFRA partnership that was clearly beset by problems at this stage of the floods.

As for the local responses to the 2007 floods, the performance of the local authorities and Local Resilience Forums (LRFs) came in for heavy criticism. According to the Pitt Review, LRF performance was hampered by a range of problems, which included a lack of established protocols between responders as well as a 'degree of confusion' as to whose responsibility it was to lead the consultation and response-activation process.[60] These 'internal' problems appear to have been compounded by a failure on the part of the LRFs to engage before the floods in 'collective' agreement on risk assessment and response over a wider, cross-boundary area. In other words, it appears that the emphasis on localism has produced a form of LRF 'parochialism' that concentrates on 'internal', geographically limited problems and neglects wider area issues. As an illustration, although the Pitt Review notes that there were many instances of effective 'mutual aid' agreements coming into effect, these were found to be highly sector-specific: 'Beyond the emergency services, however, there are few structured arrangements for mutual aid. Where it does happen it is usually ad hoc.'[61] The responses to the 2007 floods reveal that the civil protection function is hybrid in its current form. It is not a fully developed, operable, 'nodal' or heterarchical framework; neither does it resemble a straightforward command and control structure.

Based on these two acute events, what are the reasons for the quiescence of Part II? An integrated response should consider the full range of emergency powers and governance structures. But what has emerged is a picture of disdain and even neglect of Part II. Six reasons are posited for this pattern.

First is the pragmatic success of Part I alone – terrorist attacks have not been so sustained or widespread, floods have not affected such vital locations or infrastructure as to be catastrophic, and pandemic diseases have been contained. Thus, the mechanisms under Part I have worked passably well,[62] while sectoral legislation has also come into play, especially in counter-terrorism.

Second, the invocation of Part II involves costs. There is a political cost in the implicit admission that governmental planning has failed, and there are added financial costs: 'a politician who wants to raise taxes today to minimise the risk of a catastrophe of a thousand years hence will be cutting his own throat.'[63]

Third, public sentiment may be hostile to drastic state intervention and relative confidence in normal powers can serve well enough. More generally one must guard against the dangers that civil contingency efforts will divert from other social policies, that the expense will be ruinous and that the inconvenience and intrusion of security will serve as constant reminders of our insecurity. This point has been reiterated by Michael Chertoff, Secretary of the Department of Homeland Security: 'We do not want a regime in which we are so focused on risk to the exclusion of all else that we lock everything down and we destroy our country'.[64]

Fourth, the governance model of Part II is not palatable. It does not fit with preferred late modern notions of governance – it is top down and does not involve a partnership approach. To meet the challenges of risk regulation, a late modern state such as the UK must adopt styles of governance which seek to achieve influence in complex relationships, many of which reside in the private rather than public sector (over 80% of critical national infrastructure). A degree of autonomy for the actors in the network is a common feature – relationships become horizontal (or heterarchical), decentred and networked rather than hierarchical command and control relationships, with an emphasis upon shared knowledge and consensus. Part I of the CCA 2004 operates in this way, but Part II does not. In any event central government can avoid immediate pressures for achievable action in a crisis if ready powers do not exist – the current situation reduces the assumption of political responsibility and allows for detachment and the shifting of potential blame.

Fifth, Part II is not practicable. Unlike in the US, it is not a trigger for the grant of resources,[65] since the Bellwin Scheme of emergency financial assistance to local authorities is not linked.[66] And nowadays the UK military does not have the manpower and may not have the expertise to help decisively, as already considered.

Sixth, there may be some disinclination to comply with the mechanisms for limitation and accountability in Part II. This factor will be considered in the next section as being a driver towards the creation of alternative mechanisms of response.

What has taken the place of the CCA 2004 Part II?

The accusation of a 'disinclination' to comply with the mechanisms for limitation and accountability in Part II implies an underlying recognition of the need for powers like those in Part II. If that is a correct analysis, then provisions should have emerged to supplant Part II with alternatives which are easier for the executive to handle. Two measures might be claimed to fall into this category.

Draft Detention of Terrorist Suspects (Temporary Extension) Bills 2011

The draft Detention of Terrorist Suspects (Temporary Extension) Bills 2011 follow the decision of the coalition government to reduce the maximum period of detention for terrorist suspects under the Terrorism Act 2000 from 28 days to 14 days.[67] This decision was reflected in the Terrorism Act 2006 (Disapplication of Section 25) Order 2010,[68] and was then confirmed by Lord Macdonald's *Review of Counter Terrorism and Security Powers*,[69] which endorsed the views expressed in the Home Office's internal *Review of Counter Terrorism and Security Powers*.[70] The changes became permanent under the Protection of Freedoms Act 2012, sections 57 and 58.

But there was always a nagging doubt about this liberalising measure. The balance of opinion in the foregoing Reviews and in subsequent debates (including by the Parliamentary Joint Committee on the Draft Detention of Terrorist Suspects (Temporary Extension) Bills in 2011[71]) was that a maximum period of 14 days is adequate save in exceptional circumstances. There had in fact been no need to hold any terrorist suspect without charge for more than 14 days since 2007. But the 2011 reviews and the Joint Committee all concluded that the possibility of pre-charge detention beyond 14 days cannot be excluded.[72] Therefore, the Home Secretary, the police, and the Crown Prosecution Service expressed the view that it would be irresponsible not to provide for a power to extend the maximum period beyond 14 days in truly exceptional circumstances.[73] The Home Office therefore issued two Draft Bills – one to operate before the Protection of Freedoms Act 2012 came into force and one thereafter. Each draft bill (if ever tabled and passed) has the same effect: to extend the maximum period available for pre-charge detention from 14 days to 28 days if and when necessary. Both draft Bills are subject to a sunset clause so that the extended maximum detention power would only continue for three months from Royal Assent.

The draft Bills are not exactly the same as Part II of the CCA 2004. They more closely resemble the previous mechanism for invoking 28-day detention under the Terrorism Act 2006, which involved the issuing of a statutory order under section 25 of that Act for one year at a time. The government's reasoning is that a contingency power to extend the maximum period for pre-charge detention should be provided by primary legislation, so that the need for, and the details of, the power can be fully subject to parliamentary scrutiny. The Joint Committee disagreed with the draft Bill mechanism because it feared that parliamentary scrutiny of primary legislation would be so circumscribed in an emergency and when cases might be *sub judice* that the legislative process would be unsatisfactory and ineffective.[74] Subject to some tweaking in the Protection of Freedom Act 2012 to deal with parliamentary recesses, these criticisms were not accepted, and the surviving draft Bill now serves as the contingency mechanism.

It may be questioned why this mechanism was preferred to Part II of the CCA 2004. Part II can be invoked in the circumstances of a defined emergency within section 19, which expressly includes a terrorism emergency and can arise where 'an emergency has occurred, is occurring or is about to occur' (section 21(2)) – thus, the CCA 2004 can be proactive as well as reactive. The Part II definition of emergency contends against intervention in individual cases, since emergency requires a degree of scale – it 'threatens serious damage to the security of the United Kingdom'. Thus most terrorist attacks cannot justify a declaration of emergency. This limit becomes a problem if, as in the view of Lord Carlile,[75] it is desired to hold open the possibility of 28-day detention for an individual case. It is not problematic if one believes, like Lord Macdonald, that 28-day detention is only conceivable in the case of a concerted set of serious terrorist attacks which overwhelm the police.[76] Another advantage of Part II is that it contains details in sections 20 and 21 as to the triggers for the powers, all of which can be reviewed in Parliament and by the courts. There are also safeguards as to parliamentary scrutiny, a sunset clause, and the promise of subsequent independent review. More safeguards could certainly be imagined and would be worthwhile within Part II. But, even as it stands, Part II offers safeguards well in advance of what is offered by the draft Bills. The CCA 2004 includes 'Triple Lock' safeguards – that restraints will be imposed on the triggering definitions by reference to seriousness, necessity and geographical proportionality. In one sense it is perplexing to be seen as encouraging the potentially widespread use of the 'Doomsday' powers in Part II. However, it should be appreciated that the exceptionality of the powers amounts to a strong safeguard. The fact that the CCA 2004 requires the establishment of an emergency is itself a powerful deterrent, setting a high bar which is ducked under by the draft Bills, which notably specify no definition of 'emergency'.

In conclusion, the government should have had the courage of its convictions in the Protection of Freedoms Act 2012, section 57, and truly brought about the 'Permanent reduction' of the detention period as it was termed in the original Bill.[77] It should have simply abolished section 25 without feeling the need to wave the shroud of the draft Bills. If reassurance is nevertheless felt necessary for faint hearts, it should arise through fuller contingency planning under the CCA 2004. But the government wanted a 'softer' emergency power than Part II.

Enhanced Terrorism Prevention and Investigation Measures

Another illustration of the supplanting of Part II, very much with the same progeny and characteristics, concerns the draft Bill for Enhanced Terrorism Prevention and Investigation Measures (ETPIMs).[78]

The story starts again with the determination of the coalition government to liberalise aspects of the counter-terrorism laws, in this case control orders under the Prevention of Terrorism Act 2005, which were issued by the Home Secretary to restrict the freedoms of suspected terrorist by measures such as house curfews and relocation.[79] The same process of the Home Office's *Review of Counter-Terrorism and Security Powers*, published alongside Lord Macdonald's accompanying observations resulted in the repeal of control orders by the Terrorism Prevention and Investigation Measures Act 2011.[80] However, that Act replaces one executive device with another – the Terrorism Prevention and Investigations Measures (TPIMs). Like control orders, TPIMs remain in essence executive measures designed to curtail the activities of individuals suspected of involvement in terrorism, the risk from whom, the government argues, cannot be sufficiently allayed by prosecution, deportation or security surveillance.

Despite their many similarities, TPIMs may be considered a weaker mechanism than control orders. They ended the relocation power and various other restrictions at the harsher end of the scale, although most features which made control orders unpalatable to critics do continue. This relative weakness was picked up by opposition politicians who warned about 'watering down measures proven to prevent terrorist activity'.[81] In addition, the Home Office's *Review* (but not the *Macdonald Review*) avowed the possibility of ETPIMs during exceptional circumstances.[82] But it is perhaps telling that the draft Bill was not published until 1 September 2011,[83] long after the publication of the TPIM Bill, suggesting some afterthought about the need for reassurance more for the benefit of security professionals than for opposition politicians.[84] The minister was keen to emphasise that this extra platform did not represent 'some sort of U-turn',[85] but the compromise is clear. Assuming the triggers are met, an ETPIM can be more severe than a standard TPIM in several respects, including: a relocation requirement; a requirement that the individual remain in their residence for a specified period, not necessarily confined to 'overnight' and therefore reverting to a curfew; a geographical boundary restriction by which the person may not leave a specified area; a total prohibition on the possession or use of electronic communications devices or on financial services; and a wider restriction on association or communication with other persons.

The draft Bill allows for ETPIMs during 'exceptional circumstances', but these 'exceptional circumstances' are again not defined. The minister ventured the following scenarios: 'credible reporting could point to a series of concurrent attack plots, all of which appear imminent, or it might apply in the wake of a major terrorist attack when there is the prospect of further attacks to follow'.[86] A related uncertainty affecting the draft Bill device is its relationship to derogation under the European Convention on Human Rights article 15. The Home Office claim is that derogation is not a necessary condition.[87] That derogation is not ruled out contrasts with earlier promises.[88]

Another aspect of the official justification for ETPIMs was that the CCA 2004 was said to be inappropriate for three reasons: there is no explicit powers to issue TPIMs in the CCA (but equally no denial that its powers are wide enough to do so); the maximum penalty for any offence involving breach of CCA regulations is just three months' imprisonment; and the CCA triggers for invocation are too restrictive.[89] It is suggested that the last point is the most fundamental. The government has preferred once again to create a new form of 'soft' emergency in preference to the 'hard' emergency of the CCA 2004, thereby evading the carefully constructed safeguards of the latter. If the draft Bill is ever triggered, ETPIMs will remain in force for one year, extendable by one year after consultation with the Independent Reviewer of Terrorism Legislation, the Intelligence Services Commissioner and the Director General of the Security Service (cl.9(3)) but not the police or Director of Public Prosecutions.

Although the government eventually expressed a desire for pre-legislative scrutiny,[90] none of these arguments or criticisms was effectively scrutinised in Parliament during the passage of the TPIM Bill, save that sections 26 and 27 were inserted to deal with any period when Parliament is dissolved, during which a Bill could not be introduced. This potential gap was identified during pre-legislative scrutiny of the Draft Detention of Terrorist Suspects (Temporary Extension) Bill to extend periods of pre-charge detention for terrorist suspects to 28 days. After the passage of the TPIM Act, a Joint Committee on the Draft Enhanced Terrorism Prevention and Investigation Measures Bill was established and reported in November 2012.[91] The Committee was very doubtful about the efficacy of the draft Bill mechanism, especially because there is no definition of 'emergency' and no real opportunity to question the factual necessity if ever invoked. It suggested there should be private consultation, for example with the Intelligence and Security Committee.[92] On balance, it preferred an order-making power to either the draft Bill device or the use of the CCA 2004.[93]

In response, the government refused to insert a definition of 'exceptional circumstances', although it did describe them as:

> a situation where there was credible reporting pointing to a series of concurrent, imminent attack plots, or the period immediately following a major terrorist incident where we faced the prospect of further attacks. A decision to enact the ETPIM legislation is unlikely to be triggered solely by a change to the overall terrorism threat level in the absence of other factors.[94]

It also gave no undertaking to hold prior discussions with the Intelligence and Security Committee.[95] It stood by its decisions not to include the measures within the TPIM Act 2011, giving as the reason that 'these more stringent powers should only be on the statute book in exceptional circumstances'.[96] This abhorrence of sullying the statute book with the envisioning of distasteful eventualities is not, however, used as an argument for the repeal of Part II of the CCA 2004 or other legislation based on the notion of emergency, such as the Energy Act 1976, section 3, the Railways Act 1993, section 118, the Animal Health Act 2002, the Food and Environment Protection Act 1985, Part I, and a host of other examples which all specifically refer to action allowed or required in 'emergency' circumstances.

Critique

Laying a draft Bill to deal with a future emergency was said to be 'deeply disorderly and shambolic' by the Shadow Home Secretary during the TPIM debates.[97] But she perhaps forgot that the precedent for doing so was set by her own Labour Home Secretary predecessor, Jacqui Smith, who produced such a draft Bill – the Counter Terrorism (Temporary Provisions) Bill 2008 – in a fit of pique upon her defeat in October 2008 over the proposal for 42-day detention during the Counter-Terrorism Bill 2007–08.[98]

A draft Bill is preferable to an unveiling during the febrile atmosphere of a crisis, but the device only postpones, rather than avoids, the evils of emergency legislation which were well indicated by the House of Lords Select Committee on the Constitution.[99] The Coalition Government was not convincing with its argument that 'We do not believe that it is necessary to have these additional measures in the TPIM Bill as we sincerely hope that they will never be required'.[100] The measures are evidently worth the trouble of drafting and tabling. The enhanced measures should either have been incorporated in the main TPIM Bill (but not brought into force without a further order), or the CCA 2004 should have been preferred to this 'soft' emergency legislation.

Conclusions: whither Part II?

Overall, the draft Detention of Terrorist Suspects (Temporary Extension) Bills and the draft ETPIM Bill demonstrate that Part II of the Civil Contingencies Act 2004 is considered to be an irrelevance – it has become little more than a symbolic assertion of government power. This message is also affirmed by the way Part II has been elaborated since 2004, the reluctance to consider its invocation, and the underlying reasons which create resistance to it.

This outcome is both good and bad. It is good in that some positive reasons underline the disuse. These include the improved performance of planning and resilience under Part I of the CCA 2004. The design and delivery of sectoral legislation is also important. But the outcome is undesirable in the sense that the driving rationale for Part II remains – that there is a need to work out, explain and practice the responses to contingencies at all levels. There remains the possibility of national or pandemic crisis, and recent evidence, especially that of the 7/7 bombs and the 2007 floods, suggests that not all the required mechanisms are yet in place.

The final reflection, looking to the future, is that, if Part II is too blunt for practical use, then perhaps it is time to look systematically at whether a comprehensive 'soft emergency' law would be worthwhile. Such a law might pick out a range of topics (which probably already lurk in Cabinet Office and Home Office filing systems) and build on the draft Bills experiences. The results would not be as effective as Part II in terms of subject coverage or the observance of constitutional normative standards such as executive restraint. But the acquiescence of Parliament in this new executive trend of 'soft' emergencies suggests that some legislation would be better able than a dead letter Part II to secure the dynamic qualities necessary of civil contingencies laws being powerful, influential, pervasive, and commanding control over executive dominance in the most unforgiving of circumstances.

Acknowledgement

This paper draws upon materials prepared with Jim Broderick for a conference organised by the *Guardian*, 'Public Sector Emergency Planning and Business Continuity', held in London, 2007.

Notes

1. Twentieth Century-Fox Film Corporation, 2004.
2. See Clive Walker and James Broderick, *The Civil Contingencies Act 2004: Risk, Resilience and the Law in the UK* (Oxford: Oxford University Press, 2006); James Broderick, 'Intelligence and Civil Protection in the UK: Assimilating "Risky" Information in an Uncertain World', *Democracy and Security* 3 (2007), 1; and Frank Gregory, 'The UK Draft Civil Contingencies Bill and the Subsequent Act', in *Homeland Security in the UK*, ed. P. Wilkinson (Abingdon: Routledge, 2007).
3. For details of the manifold pieces of legislation, see Clive Walker, *Terrorism and the Law* (Oxford: Oxford University Press, 2011).

4. See The Pitt Review, *Learning Lessons from the 2007 Floods*, http://webarchive.nationalarchives.gov.uk/20100807034701/http:/archive.cabinetoffice.gov.uk/pittreview/thepittreview/final_report.html (accessed November 10, 2013); and Flood and Water Management Act 2010.
5. See *Foot and Mouth Disease 2001: The Lessons to be Learned Inquiry by Dr Iain Anderson* (2001–02 HC 888); *Response to the Reports of the Foot and Mouth Disease Inquiries*, Cm. 5637, (London, 2002); and Animal Health Act 2002.
6. Deirdre Hine, *The 2009 Influenza Pandemic: An Independent Review of the UK Response to the 2009 Influenza Pandemic* (London: Cabinet Office, 2010).
7. Frank Furedi, *Culture of Fear: Risk Taking and the Morality of Low Expectation* (London: Continuum, 2002).
8. For histories, see Walker and Broderick, *The Civil Contingencies Act 2004*, chap. 2.
9. See Phil Hathaway, *The Effect of Fuel Protest on Road Traffic* (London: Transport Statistics Road Traffic Division, DETR, 2000). The event is considered to have been the closest to triggering what are now Part II powers in recent years. *Hansard* House of Commons Standing Committee F, February 2004, col.147 3, Fiona Mactaggart.
10. For full details, see Walker and Broderick, *The Civil Contingencies Act 2004*, chap. 8.
11. Cabinet Office, *Emergency Response and Recovery*, 5th ed. (London, 2013), https://www.gov.uk/government/publications/emergency-response-and-recovery.
12. A Government Liaison Officer may be assigned to work alongside the SCG. Cabinet Office, *Emergency Response and Recovery*, para 9.1.2.
13. Ibid., paras 4.2.25. See also Cabinet Office, 'National Recovery Guidance', https://www.gov.uk/national-recovery-guidance (accessed December 11, 2013); and Cabinet Office, *Responding to Emergencies*: *The UK Central Government Response – Concept of Operations* (London, 2013): para 5.6.
14. See Department for Culture, Media and Sport (DCMS), *A Detailed Guide to Roles and Responsibilities in Humanitarian Assistance* (London, 2006), https://www.gov.uk/government/uploads/system/uploads/attachment_data/file/61223/ha_rolesandresponsibilities.pdf; Cabinet Office, *Identifying People Who are Vulnerable in a Crisis*: *Guidance for Emergency Planners and Responders* (London, 2008), https://www.gov.uk/government/uploads/system/uploads/attachment_data/file/61225/vulnerable_guidance_0.pdf; Cabinet Office, *Humanitarian Assistance in Emergencies*: *Non-Statutory Guidance on Establishing Humanitarian Assistance Centres* (London, 2011), https://www.gov.uk/government/uploads/system/uploads/attachment_data/file/61221/hac_guidance.pdf; and DCMS, *Humanitarian Assistance Strategic Guidance*: *Building Capability to Look after People Affected by Emergencies* (London, 2011), https://www.gov.uk/government/uploads/system/uploads/attachment_data/file/61220/final-humanitarian-assistance-guidance.pdf. For disasters abroad, the Foreign & Commonwealth Office can provide assistance through a Rapid Deployment Team network, London Crisis Response Teams, and a Crisis Centre, as well as diplomatic assistance. See https://www.gov.uk/government/policies/supporting-british-nationals-overseas.
15. Cabinet Office, *Emergency Response and Recovery*, para 13.2.
16. Cabinet Office, *The Lead Government Department and its Role – Guidance and Best Practice* (London, 2004), https://www.gov.uk/government/uploads/system/uploads/attachment_data/file/61355/lead-government-departments-role.pdf, chap. 1.
17. Cabinet Office, *Emergency Response and Recovery*, para 13.4.
18. Ibid., 187.
19. Cabinet Office, *Responding to Emergencies*, para 1.7.
20. See Lady Justice Hallett, *Coroner's Inquests into the London Bombings of 7 July 2005: Report under Rule 43 of The Coroner's Rules 1984* (London: Ministry of Justice 2011); and Cabinet Office, *Coroner's Inquests into the London Bombings of 7 July 2005*: *Review of Progress* (London, 2012).
21. See Mark Sandford, *The Abolition of Regional Government*, SN/PC/05842 (London: House of Commons, 2011).
22. Cabinet Office, *Emergency Response and Recovery*, para 1.6.
23. Cabinet Office, *Emergency Preparedness* (London, 2006–12), https://www.gov.uk/government/publications/emergency-preparedness. The current edition has 591 pages.
24. Cabinet Office, *Emergency Response and Recovery*, paras 14.1.1, 14.3.8.
25. See, further, Michael Head, 'Calling out the Troops and the Civil Contingencies Act', *Public Law* [2010]: 340.

26. See, further, Ministry of Defence (MoD), *JDP 02: Operations in the UK – The Defence Contributions to Resilience*, 2nd ed. (London, 2007), Annex 2D, Annex 2E, https://www.gov.uk/government/publications/operations-in-the-uk-a-joint-doctrine-publication.
27. MoD, *Operations in the UK*. A shorter version has been issued to civil responders. *JDP 02: Addendum*, 2010, https://www.gov.uk/government/publications/jdp-02-addendum-to-operations-in-the-uk-the-defence-contribution-to-resilience. It likewise states (para 24a) that 'Military aid should only be provided where the need for someone to act is clear and where other options have been discounted by the Civil Responder. The use of mutual aid, other agencies, and the private sector must be otherwise considered as insufficient or be unsuitable.'
28. MoD, *Operations in the UK*, para 205.
29. Cabinet Office, *Emergency Preparedness: Guidance on Part 1* (London, 2011), para 15.5, https://www.gov.uk/government/uploads/system/uploads/attachment_data/file/61038/Chapter-15-other-sectors-involved-in-EP-amends-10112011.pdf.
30. See Christopher Whelan, 'Military intervention in industrial disputes', *Industrial Relations Journal* 18 (1979): 222; Steve Peak, *Troops in Strikes* (London: Cobden Trust, 1984), chap. 2; Peter J. Rowe and Christopher J. Whelan, eds., *Military Intervention in Democratic Societies* (London: Croom Helm, 1985); Gillian S. Morris, *Strikes in Essential Services* (London: Mansell, 1986), chap. 4; and Anthony Babington, *Military Intervention in Britain* (London: Routledge, 1990).
31. Ministry of Defence, *The Strategic Defence Review: A New Chapter*, Cm. 5566 (London, 2002), chap. 5.
32. Ministry of Defence, *Operations in the UK*, Annex 2F para 2F5.
33. *Hansard* (House of Commons), vol. 489 col. 1302 W, 19 March 2009, Bob Ainsworth.
34. *Defence and Security in the UK* (2001–02 HC 518), para 226.
35. *Hansard*, vol. 421 col. 1362, 24 May 2004, Hazel Blears.
36. See, further, Leonard Jason-Lloyd, 'Emergency Workers (Obstruction) 2006 – an overview', *Justice of the Peace* 171 (2007): 174.
37. Compare Emergency Workers (Scotland) 2005, which also forbids assaults and applies additionally to health workers, but note the Sentencing Guidelines Council document, *Overarching Principles: Seriousness* (London, 2004), para 1.23: 'Factors indicating a more than usually serious degree of harm ... Offence is committed against those working in the public sector or providing a service to the public'.
38. See Police Act 1996 s 89 (police constables and applied also to prison officers by the Prison Act 1952 s 8); Courts Act 2003 s 57 (court security officers); Criminal Justice Act 1991 s 90 (prison custody officers); Immigration & Asylum Act 1999 Sched. 11 paras 4, 5 (detainee custody officers); and Police Reform Act 2002 s 46 (police custody support officers).
39. See Cabinet Office, *Emergency Planning Review: The Future of Emergency Planning in England and Wales: Results of the Consultation* (London, 2002), para 6.
40. Ministry of Justice, *The Governance of Britain: War Powers and Treaties*, Cm. 7239 (London, 2007). See, further, the paper by Blick in this special issue.
41. Cabinet Office, *The Governance of Britain: Review of the Executive Royal Prerogative Powers – Final Report* (London, 2009), paras 70–72.
42. Ibid., para 75.
43. For further comment, see House of Commons Political and Constitutional Reform Committee, *Parliament's Role in Conflict Decisions* (2010–12 HC 923) and government responses at 1477, 1673; and House of Lords Select Committee on the Constitution, *Constitutional Arrangements for the Use of Armed Force* (2013–14 HL 46).
44. Home Affairs Committee, *The Home Office's response to Terrorist Attacks* (2009–10 HC 117), para 7; and *Government Reply*, Cm. 7788 (London, 2010), p 4
45. See, further, Clive Walker, 'Governance of Critical National Infrastructure', *Public Law* [2008]: 323.
46. See https://www.gov.uk/government/publications/sector-resilience-plans.
47. Only the CCA 2004 Part I is mentioned. Cabinet Office, *A Summary of the 2012 Sector Resilience Plans* (London, 2012), p. 8.
48. Cabinet Office, *Civil Contingencies Act Enhancement Programme (CCAEP): Programme Initiation Document*, Version 1.7 (London, 2010), para 13.
49. See Cabinet Office, *Update of Emergency Response and Recovery – Non-statutory Guidance to Complement Emergency Preparedness: Consultation Paper* (London, 2009).

50. Christopher Hood, Oliver James, George Jones, Colin Scott, and Tony Travers, *Regulation Inside Government: Waste-Watchers, Quality Police and Sleaze-Busters* (Oxford: Oxford University Press, 1999).
51. Michael Moran, *The British Regulatory State: High Modernism and Hyper Innovation* (Oxford: Oxford University Press, 2003), 4.
52. Civil Contingencies Act 2004 (Commencement No.1) Order 2004 SI No. 3281.
53. Home Office, *Addressing Lessons from the Emergency Response to the 7 July 2005 London Bombings* (London: Home Office, 2006), para 89.
54. See HM Government, *Countering International Terrorism: The United Kingdom's Strategy*, Cm. 6888 (London, 2006), paras 99, 100.
55. M. Pitt, *Learning Lessons from the 2007 Floods*, Interim Report (London: Cabinet Office, 2007); and Pitt, *Learning Lessons from the 2007 Floods*, Final Report (London: Cabinet Office, 2008). There is no mention of the CCA in House of Commons Environment, Food and Rural Affairs Select Committee, *Managing Flood Risk* (2013–14 HC 330).
56. Final Report, chs 12, 13.
57. Final Report, para 13.41.
58. Final Report, para 13.42, 13.43.
59. First Report, para 5.71.
60. First Report, paras 5.62, 5.63.
61. First Report, para 5.16.
62. As well as the Pitt Review assessments already mentioned, the verdict on the responses to the 2009 influenza pandemic also was broadly positive. Hine, *The 2009 Influenza Pandemic.*
63. R.A. Posner, *Catastrophe: Risk and Response* (New York: Oxford University Press, 2004), 118.
64. Quoted by D. Sevastopulo, 'US focus on risk "poses threat to economy"', *Financial Times*, March 13, 2006, 1.
65. See Walker and Broderick, *The Civil Contingencies Act 2004*, chap. 9.
66. Ibid., chap. 8; and Mark Sandford, *The Bellwin Scheme*, SN/PC/00643 (London: House of Commons Library, 2013). For the current implementation in 2013–14, see https://www.gov.uk/government/publications/bellwin-scheme-2013-to-2014-guidance.
67. Draft Detention of Terrorist Suspects (Temporary Extension) Bills, Cm. 8018 (London, 2011). See, further, Walker, *Terrorism and the Law*, chap. 5; and Ed Cape, 'The Counter-terrorism Provisions of the Protection of Freedoms Act 2012: Preventing Misuse or a Case of Smoke and Mirrors?', *Criminal Law Review* [2013]: 385.
68. Terrorism Act 2006 (Disapplication of Section 25) Order 2010, SI 2010/1909.
69. Lord MacDonald, *Review of Counter Terrorism and Security Powers*, Cm. 8003 (London, 2011).
70. Home Office, *Review of Counter Terrorism and Security Powers*, Cm. 8004 (London, 2011).
71. *Report* (2010–12 HL 161/ HC 893).
72. See also Lord Carlile, *Report on Proposed Measures for Inclusion in a Counter Terrorism Bill*, Cm. 7262 (London, 2007), para 46.
73. Joint Committee on the Draft Detention of Terrorist Suspects (Temporary Extension) Bills, *Report* (2010–12 HL 161/ HC 893), para 49.
74. Ibid., para 94.
75. Ibid., para 44.
76. Ibid., para 43.
77. No. 146 cl.57.
78. Home Office, *Draft Enhanced Terrorism Prevention and Investigation Measures Bill*, Cm. 8166 (London, 2011). See, further, D. Anderson, *Control Orders in 2011* (London: Home Office, 2012), 5.39; and Helen Fenwick, 'Designing ETPIMs around ECHR Review or Normalisation of "Preventive" Non-trial-based Executive Measures?', *Modern Law Review* 76 (2013): 876.
79. See Walker, *Terrorism and the Law*, chap. 7.
80. See Helen Fenwick, 'Preventive Anti-terrorist Strategies in the UK and ECHR: Control Orders, TPIMs and the Role of Technology', *International Review of Law, Computers & Technology* 25 (2011): 129; Tamara Tulich, 'Adversarial Intelligence? Control Orders, TPIMs and Secret Evidence in Australia and the United Kingdom', *Oxford University Commonwealth Law Journal* 12 (2012): 341; Clive Walker and Alexander Horne, 'The Terrorism Prevention

and Investigations Measures Act 2011: One Thing but Not Much the Other?', *Criminal Law Review* [2012]: 421; and Alan Hunt, 'From Control Orders to TPIMs: Variations on a Number of Themes in British Legal Responses to Terrorism', *Crime, Law & Social Change* (2013) (published online 28 September 2013).

81. *Hansard*, vol. 529 col. 74, 7 June 2011, Yvette Cooper.
82. Home Office, *Review of Counter Terrorism and Security Powers*, 43.
83. It was accompanied by what are now ss.26 and 27 of the TPIM Act, introduced at the Commons Report Stage to deal with any period when Parliament is dissolved, during which a Bill could not be introduced. *Hansard*, vol. 532 col. 87, 5 September 2011.
84. Compare the intention not to publish expressed by the Security Minister (Baroness Neville Jones) on 8 February 2011. Joint Committee on Human Rights, *Counter-Terrorism Review* (2010–11 HC 797-i), Qq51–56.
85. *Hansard*, vol. 532 col. 133, 5 September 2011, James Brokenshire.
86. Ibid., cols 105–106.
87. Home Office, *Draft Enhanced Terrorism Prevention and Investigations Measures Bill: Memorandum to the Joint Committee on Human Rights* (London, 2011), para 15.
88. Home Office, *Review of Counter Terrorism and Security Powers*, 43.
89. House of Lords Constitution Committee, *Prevention and Investigation Measures Bill Report* (2010–12 HL 198); and *Government Reply*, Letter of 27 October 2011, 5.
90. Home Office, *Draft Enhanced Terrorism Prevention and Investigation Measures Bill*, para 5.
91. *Report* (2012–13 HL 70/HC 495).
92. Ibid., para 3.
93. Ibid., para 31.
94. *Government Response*, Cm. 8536 (London, 2013), para 3.
95. Ibid., para 5.
96. Ibid., para 8.
97. *Hansard*, vol. 532 col. 135, 5 September 2011, Yvette Cooper.
98. See http://webarchive.nationalarchives.gov.uk/20090414164200/http://press.homeoffice.gov.uk/Speeches/counter-terrorism-bill-statement.
99. House of Lords Select Committee on the Constitution, *Fast-track Legislation: Constitutional Implications and Safeguards* (2008–09 HL 116).
100. *Hansard*, vol. 730 col. 1136, 5 October 2011, Lord Henley.

Terrorism and counterterrorism in the US: the question of responsible policy-making

John Mueller[a] and Mark G. Stewart[b]

[a]*Mershon Center for International Security Studies, Ohio State University, USA;* [b]*Centre for Infrastructure Performance and Reliability, University of Newcastle, Australia*

Officials serving the public are tasked at the most fundamental level to spend funds in a manner that most effectively and efficiently keeps people safe. To do otherwise is irresponsible and, because human lives are at stake, immoral. In the case of counterterrorism policy-making, it is important to evaluate the degree to which any gains in security afforded by counterterrorism measures have been great enough to justify their cost. Risk analysis is an aid to responsible decision-making that does exactly that. We deal with four issues central to this approach, applying them to the hazard presented by terrorism: the cost per saved life, acceptable risk, cost–benefit analysis, and risk communication. We also assess the degree to which risk analysis has been coherently applied to counterterrorism efforts in the US in making or evaluating decisions that have cost taxpayers many hundreds of billions of dollars over the past dozen years.

Introduction

As declared in the first sentence of the American constitution and throughout the work of Thomas Hobbes, especially *Leviathan*, a key reason for founding governments is to 'ensure domestic Tranquility'. Accordingly officials serving the public are tasked at the most fundamental level to spend funds in a manner that most effectively and efficiently keeps people safe. Doing so is neither easy nor precise, and the funds available for this purpose are, of course, limited. Moreover, there are inevitably distortions stemming from public and personal emotion and from political pressures. But to the degree possible the task should be carried out systematically and professionally. To do otherwise is irresponsible, a profound betrayal of the public trust and, because human lives are at stake, immoral.

Terrorism is a hazard to human life, and it should be dealt with in a manner similar to that applied to other hazards – albeit with an appreciation for the fact that terrorism often evokes extraordinary fear and anxiety. While allowing emotion to overwhelm sensible analysis is both understandable and common among ordinary people, it is not appropriate for officials charged with – responsible for – keeping them safe. If such officials are incapable of carrying out their jobs in a manner that provides the most public safety for the money expended, they should frankly admit they are being irresponsible by elevating public satisfaction with their performance over public safety – or they should resign from their posts. People who join the army or become fire-fighters accept the possibility that at some point they may be put in a position in which they are shot at or required to enter a burning building. People who become decision-makers should in equal measure acknowledge that, in order to carry out their job properly and responsibly, they may be required on occasion to make some difficult, unpopular and even career-threatening decisions.

Risk analysis is an aid to responsible decision-making that has been developed, codified and applied over the past few decades – or in some respects centuries.[1] In this article we deal with four issues central to this approach and apply them to the hazard presented by terrorism: the cost per saved life, acceptable risk, cost–benefit analysis and risk communication. We conclude with an assessment of the degree to which risk analysis has been coherently applied to counterterrorism efforts by government – particularly by the US government – in making or evaluating decisions that have cost taxpayers hundreds of billions of dollars over the past dozen years.

Cost per saved life

When regulators propose a new rule or regulation to enhance safety, they are routinely required to estimate how much it will cost under their proposal to save a single life. Table 1 supplies information about how this calculation comes out for dozens of government rules and regulations in the US.

The results are anything but tidy, and they often reflect psychological and political aspects of risk perception or electoral and lobbyist pressure. However, some general tendencies and limits have been established over time. Thus, looking over such data, Elizabeth Paté-Cornell suggests that a ceiling of \$3 million per life saved, inflation adjusted to 2010 dollars, seems roughly appropriate in current practice, although there are several entries in Table 1 that are substantially, even spectacularly, higher.[2] But, in general, regulators and administrators begin to become unwilling to spend more than \$1 million to save a life, and they are extremely reluctant to spend over \$10 million, preferring instead to expend funds on alternative measures that save lives at a lower cost.

This approach can be, and has been, expanded to embrace deaths by terrorism. Following the widely applied approach a study for the US government's Department of Homeland Security by Lisa Robinson and her colleagues concludes that the best estimate of a value of a saved human life for homeland security analysis would be about \$7 million in 2013 dollars.[3] Most studies focus on relatively common risks, such as workplace or motor vehicle accidents, and the Robinson study goes on to suggest that 'more involuntary, uncontrollable, and dread risks may be assigned a value that is perhaps twice that of more familiar risks', or some \$14 million. This approach essentially adds into the analysis of much of the substantial indirect and ancillary costs, including emotional ones, associated with a terrorist event.

The US spends about \$100 billion per year on seeking to deter, disrupt or protect against domestic counterterrorism. If each saved life is valued at \$14 million, it would be necessary for the counterterrorism measures to prevent or protect against between 6000 and 7000 terrorism deaths in the county each year or twice that if the lower figure of \$7 million for a saved life is applied.

These figures seem very high. The total number of people killed by Islamist extremist terrorists within the US since 9/11 is 19, or fewer than two per year, a far cry, of course, from

Table 1. Regulatory expenditure per life saved

Regulation	Year	Agency	Cost per life saved in 2010 dollars
Steering column protection standards	1967	NHTSA	140,000
Unvented space heater ban	1980	CPSC	140,000
Seatbelt/air bag	1984	NHTSA	140,000
Aircraft cabin fire protection standard	1985	FAA	140,000
Underground construction standards	1989	OSHA	140,000
Auto fuel system integrity	1975	NHTSA	710,000
Trihalomethane in drinking water	1979	EPA	850,000
Aircraft seat cushion flammability	1984	FAA	850,000
Alcohol and drug controls	1985	FRA	850,000
Aircraft floor emergency lighting	1984	FAA	990,000
Concrete and masonry construction	1988	OSHA	990,000
Passive restraints for trucks and buses	1989	NHTSA	1,100,000
Children's sleepwear flammability ban	1973	CPSC	1,400,000
Auto side impact standards	1990	NHTSA	1,400,000
Metal mine electrical equipment standards	1970	MSHA	2,400,000
Trenching and evacuation standards	1989	OSHA	2,600,000
Hazard communication standard	1983	OSHA	2,700,000
Trucks, buses and MPV side-impact	1989	NHTSA	3,700,000
Grain dust explosion prevention	1987	OSHA	4,700,000
Rear lap/shoulder belts for autos	1989	NHTSA	5,400,000
Standards for radionuclides in uranium mines	1984	EPA	5,800,000
Ethylene dibromide in drinking water	1991	EPA	9,700,000
Asbestos occupational exposure limit	1972	OSHA	14,000,000
Benzene occupational exposure limit	1987	OSHA	15,000,000
Electrical equipment in coal mines	1970	MSHA	15,800,000
Arsenic emission standards for glass plants	1986	EPA	22,900,000
Cover/move uranium mill tailings	1983	EPA	76,100,000
Acrylonitrate occupational exposure limit	1978	OSHA	87,000,000
Coke ovens occupational exposure limit	1976	OSHA	107,400,000
Arsenic occupational exposure limit	1978	OSHA	180,800,000
Asbestos ban	1989	EPA	187,200,000
1,2-Dechloropropane in drinking water	1991	EPA	1,103,900,000
Hazardous waste land disposal ban	1988	EPA	7,084,000,000
Municipal solid waste landfills	1988	EPA	32,300,000,000
Formaldehyde occupational exposure limit	1987	OSHA	145,723,000,000
Atrazine/alachlor in drinking water	1991	EPA	155,640,000,000
Hazardous waste listing for wood-preserving chemicals	1990	EPA	9,635,870,000,000

Source: Adapted by Mark Stewart from W. Kip Viscusi, 'The Value of Life in Legal Contexts: Survey and Critique', *American Law and Economics Review* 2, no. 1 (2000): 195–222.

6000 to 7000 per year.[4] A defender of the expenditure on terrorism might argue that the number is that low primarily because of the counterterrorism efforts. Others might be more sceptical.

Acceptable risk

Another way to approach the issue is to compare the annual fatality rates caused by terrorism with those caused by other hazards. Table 2 provides relevant information. It leads to a consideration of the central analytic issue of acceptable risk. Is the likelihood of being killed

Table 2. Comparison of annual fatality risks

Hazard	Territory	Period	Total fatalities for the period	Annual fatality risk
World War II	Worldwide	1939–45	61,000,000	1 in 221
Cancers	US	2009	560,000	1 in 540
War (civilians)	Iraq	2003–08	113,616	1 in 1150
All accidents	US	2007	119,000	1 in 2500
Traffic accidents	US	2008	37,261	1 in 8200
Traffic accidents	Canada	2008	2431	1 in 13,500
Traffic accidents	Australia	2008	1466	1 in 15,000
Homicide	US	2006	14,180	1 in 22,000
Traffic accidents	UK	2008	2538	1 in 23,000
Terrorism	**Northern Ireland**	**1970–2012**	**1758**	**1 in 49,000**
Industrial accidents	US	2007	5657	1 in 53,000
Homicide	Canada	2008	611	1 in 55,000
Intifada	**Israel**	**2000–06**	**553**	**1 in 72,000**
Homicide	UK	2008	887	1 in 67,000
Homicide	Australia	2008	290	1 in 76,000
Terrorism	**US**	**2001**	**2982**	**1 in 101,000**
Natural disasters	US	1999–2008	6294	1 in 480,000
Drowning in bathtub	US	2003	320	1 in 950,000
Terrorism	**UK**	**1970–2012**	**2196**	**1 in 1,200,000**
Home appliances	US	yearly average	200	1 in 1,500,000
Deer accidents	US	2006	150	1 in 2,000,000
Commercial aviation	US	yearly average	130	1 in 2,300,000
Terrorism	**US**	**1970–2012**	**3309**	**1 in 4,000,000**
Terrorism	**Canada**	**1970–2012**	**336**	**1 in 4,300,000**
Terrorism	**UK**	**1970–2012**	**434**	**1 in 5,900,000**
Peanut allergies	US	yearly average	50–100	1 in 6,000,000
Lightning	US	1999–2008	424	1 in 7,000,000
Terrorism	**Australia incl. Bali**	**1970–2012**	**117**	**1 in 8,000,000**
Transnational terrorism	**World outside war zones**	**1975–2003**	**13,971**	**1 in 12,500,000**

Source: Terrorism fatalities sourced from the Global Terrorism Database developed by the US National Consortium for the Study of Terrorism and Responses to Terrorism (START). It contains country-by-country information for more than 80,000 terrorist incidents that have taken place throughout the world between 1970 and 2012. See http://www.start.umd.edu/gtd/.

by the hazard unacceptably high, or is it low enough to be acceptable? That is, just how safe is safe enough?

We often say that there is nothing more important than the value of human life. Yet, obviously, we don't really believe this. For example, Americans are clearly willing to sacrifice tens of thousands of lives per year to have the car, even though it is quite possible to move people without killing them: passengers killed on railways in a year can often be numbered on the fingers of one hand. Many other social policies involve the same sort of consideration. To take an extreme example, every year a few thousand people in the US die in falls from buildings that are more than one storey high. Those lives could be saved by closing off all buildings at the ground floor. To reject such a policy is to say tall buildings are worth that cost in lives. As a society, then, we regularly and inescapably adopt policies in which human lives are part of the price.

A review of 132 US federal government regulatory decisions associated with public exposure to environmental carcinogens found that regulatory action never occurred if the individual annual fatality risk was lower than 1 in 700,000.[5] Established regulatory practices in several developed countries are similar. In general risks are deemed unacceptable if the annual fatality risk is higher than 1 in 10,000 or perhaps higher than 1 in 100,000. They are deemed acceptable if the annual fatality risk is lower than 1 in 700,000 or perhaps one in a million or one in two million. Between these two regions is an area that might be considered tolerable risk.

These considerations, substantially accepted for years, even decades, by public regulatory agencies after extensive evaluation and considerable debate and public discussion, are designed to provide a viable, if somewhat rough, guideline for public policy. Clearly hazards that fall in the unacceptable range (traffic accidents, for example) should generally command the most attention and the most resources. Those in the tolerable range may also be worthy of consideration, though obviously the urgency is less, and only relatively inexpensive measures to further reduce the risk should be pursued. Those hazards in the acceptable range (drowning in bathtubs, for example) would generally be deemed of little or even negligible concern – they are risks we can live with – and further precautions would scarcely be worth pursuing unless they are quite remarkably inexpensive.

Overall, then, it is clear that governments have been able to set out, and agree upon, risk acceptance criteria for use in decision-making for a wide variety of hazards, including ones that are highly controversial and emotive, such as pollution, nuclear and chemical power plant accidents, and public exposure to nuclear radiation and environmental carcinogens.

As can be seen in Table 2, almost all annual terrorism fatality risks are lower than one in a million — for the USA, UK, Canada and Australia they are less than one in four million per year. Therefore they generally lie within the range deemed by regulators internationally to be safe or acceptable and do not require further regulation.[6] Applying conventional standards, then, terrorism currently presents a threat to human life in the Western world that is, in general, acceptable, and efforts, particularly expensive ones, to further reduce its likelihood or consequences are scarcely justified.

It is possible that any relaxation in security measures will increase the terrorism hazard risk. However, for the terrorism risk to border on becoming unacceptable by established risk conventions – that is, to reach an annual fatality rate of 1 in 100,000 – the number of fatalities from all forms of terrorism in the US and Canada would have to increase 40-fold, in Britain (excluding Northern Ireland) more than 55-fold and in Australia more than 80-fold.

We have been using historical data on terrorism here, and there is, of course, no guarantee that the frequencies of the past will necessarily persist into the future. However, there seems to be little evidence terrorists are becoming any more destructive, particularly in the West. In fact, if anything, there seems to be a diminishing, not expanding, level of terrorist activity and destruction, at least outside of war zones.[7] Those who wish to discount such arguments and projections need to demonstrate why they think terrorists will suddenly get their act together and inflict massively increased violence, visiting savage discontinuities on the historical data series.

Cost–benefit analysis

Cost–benefit analysis brings risks and costs together. A conventional approach to cost-effectiveness compares the costs of a security measure with its benefits, as tallied in lives saved and damage averted. The benefit of a security measure is a multiplicative composite of three considerations: the probability of a successful attack, absent the security measure, the losses sustained in a successful attack, and the reduction in risk furnished by the security measure. In this way: benefit = (probability of a successful attack) × (losses sustained in the attack) × (reduction in risk).

The interaction of these variables can perhaps be seen in the following example. Suppose there is a dangerous curve on a road that results in an accident from time to time. To evaluate measures designed to deal with this problem, the analyst would need to estimate: (1) the probability of an accident each year under present conditions; (2) the consequences of the accident (death, injury, property damage); and (3) the degree to which a proposed safety measure lowers the probability of an accident (erecting warning signs) and/or the losses sustained in the accident (erecting a crash barrier). If the benefit of the risk-reduction measure – these three items multiplied together – outweighs its cost, the measure would be deemed to be cost-effective.

These considerations can be usefully finessed in a procedure known as 'break-even analysis'. In this, we seek to determine what the probability of an otherwise successful terrorist attack would have to be for a security measure to begin to justify its cost. Thus we set the cost of the measure equal to its benefit:

cost = (probability of a successful attack) × (losses sustained in the attack) × (reduction in risk)

which becomes

(probability of a successful attack) = (cost) / [(losses sustained in the attack) × (reduction in risk)]

We have applied this approach to the overall increase in domestic homeland security spending in the US by the federal government (including for national intelligence) and by state and local governments. That is, we assume homeland security measures in place before the 9/11 attacks continue, and we evaluate the cost-effectiveness of the additional funds that have been allocated to homeland security. By 2009 this increase totalled some $75 billion per year. This is a very conservative measure of the degree to which homeland security expenditures have risen since 9/11, because it excludes such items as private sector expenditures, hidden and indirect costs of implementing security-related regulations and the costs of the terror-related (or terror-impelled) wars in Iraq and Afghanistan. To evaluate the reduction in risk provided by this array of security measures, we need to consider their effectiveness in deterring, disrupting or protecting against a terrorist attack.

In assessing risk reduction it is important, first, to assess the risk-reduction effectiveness of homeland security measures that were in place before 9/11. In addition, it should be kept in mind that the tragic events of 9/11 massively heightened the awareness of the public of the threat of terrorism, resulting in extra vigilance that has often led to the arrest of terrorists or the foiling of terrorist attempts at little or no cost to the government. In our analysis we assume that risk reduction caused by the security measures in place before 9/11 and by the extra vigilance of the public after that event together reduced risk by 50%. This is an exceedingly conservative estimate, because security measures that are at once effective and relatively inexpensive are generally the first to be implemented – for example, one erects warning signs on a potentially dangerous curve in the road before rebuilding the highway. Furthermore, most terrorists (or would-be terrorists) do not show much intelligence, cleverness, resourcefulness or initiative.[8] Therefore measures to deal with them are relatively inexpensive and are likely to be instituted first. Dealing with the smarter and more capable terrorists is more difficult and expensive, but these people represent, it certainly appears, a decided minority among terrorists.

For our analysis we assume that the increase in US expenditures on homeland security since 2001 have been dramatically effective, reducing the remaining risk by an additional 45%. Total risk reduction is generously assumed, then, to be 95%, with the pre-existing measures and the extra public vigilance responsible for 50% of the risk reduction and the enhanced expenditures responsible for the remaining 45%.

Putting this all together, we find that, in order for enhanced expenditures on homeland security to be deemed cost-effective under our approach – which substantially biases the consideration towards finding them effective – each year they would have to deter, disrupt or protect against 1667 otherwise successful attacks (e.g. something like the one attempted on Times Square in 2010[9]), which is more than four per day.[10]

There are extreme scenarios that can be taken to suggest that enhanced US security expenditures could be cost-effective, such as in the nightmare vision of a nuclear attack in a crowded city, as well as costly massive overreaction. However, for those who find such outcomes dangerously likely, the policy response would logically be to spend on reducing the risk of nuclear terrorism in the one case and to develop strictures to overreaction in the other. It would not be, for example, to spend tens of billions of dollars each year on protection measures.

Our findings dealing with the total enhanced homeland security expenditures should not be taken to suggest that *all* security measures necessarily fail to be cost-effective: there may be specific measures that are cost-effective. But each should be subjected to the kind of risk analysis we have applied to the overall increases in expenditure.

We have done so for several specific measures. It appears, for example, that the protection of a standard office-type building would be cost-effective only if the likelihood of a sizeable terrorist attack on the building is a thousand times greater than it is at present. Something similar holds for the protection of bridges. On the other hand, hardening cockpit doors on aircraft appears to be cost-effective. However, the provision for air marshals on the planes decidedly is not, and the cost-effectiveness of full-body scanners is questionable at best.[11] Overall, by far the most cost-effective counterterrorism measure is to refrain from overreacting.

Risk communication

Officials who seek to expend limited funds in a manner that best enhances public safety should be risk-neutral: insofar as this can be determined, they should deal with the objective likelihood the hazard will occur, and rely on that in their decision-making. Although we understand that people are often more risk-averse when considering issues like terrorism, we follow the US Office of Management and Budget requirement that governments expending tax money in a responsible manner need to be neutral when assessing risks, something that entails focusing primarily on mean estimates in risk and cost–benefit calculations, not primarily on worst-case or pessimistic ones.[12]

The willingness to accept risk, however, is influenced not only by its objective likelihood but by a variety of psychological, social, cultural and institutional processes that depend on such qualities as the uncontrollability of the risks; the dread (or fear) they inspire; their involuntary nature or catastrophic potential; whether they can be preventively controlled, are certain to be fatal, can easily be reduced, result in an inequitable distribution of risk, threaten future generations, or affect one personally; whether they are increasing or not observable, unknown to those exposed, new or unfamiliar, and unknown to science; and whether they have immediate effect or affect a large number of people.[13] It is important, then, for officials to communicate risk objectively. If they can convince their constituents to adopt a risk-neutral perspective, they will be in a far better position to expend pubic funds in ways that most enhance public safety.

Because people are often more risk-acceptant or risk-averse than an objective analysis would dictate, politicians and bureaucrats face considerable political pressure over the terrorism issue. Their dilemma is nicely parsed by James Fallows. He points out that 'the political incentives here work only one way'. A politician who supports more extravagant counterterrorism measures 'can never be proven wrong' because an absence of attacks shows that the 'measures have "worked"', while a new attack shows that we 'must go farther still'. Conversely, a politician seeking to limit expenditure 'can never be proven "right"', while 'any future attack will always and forever be that politician's "fault"'.[14]

However, it is possible that politicians and bureaucrats are overly fearful about the political consequences. Indeed, sometimes leaders have been able to restrain their instinct to overreact, and this has often proved to be entirely acceptable politically. The US did not massively overreact to terrorist bombings against its soldiers and citizens in Lebanon in 1983 or over Lockerbie, Scotland, in 1988.[15] This is a particularly important issue because, as noted, it certainly appears that avoiding overreaction is by far the most cost-effective counterterrorism measure.

Moreover, although political pressures may force actions and expenditures that are unwise, they usually do not precisely dictate the level of expenditure. Thus, although there are recurrent public demands to 'do something' about terrorism, nothing in those demands specifically requires American officials to impose the removal of shoes in airport security lines, to require passports to enter Canada, to spread bollards like dandelions, or to make a huge number of buildings into forbidding fortresses. The UK, which faces an internal threat from terrorism that seems considerably greater than that for the US, appears nonetheless to spend proportionately much less than half as much on homeland security, and the same holds for Canada and Australia.[16] Yet politicians and bureaucrats there do not seem to suffer threats to their positions, or other political problems because of this relative restraint.

It is true that few voters spend a great amount of time following the ins and outs of policy issues and even fewer are certifiable policy wonks. But they *are* grown-ups, and it is just possible that they would respond reasonably to an adult conversation about terrorism.[17]

Responsible counterterrorism policy-making

In seeking to evaluate the effectiveness of the massive increases in homeland security expenditures since September 11, 2001, the common and urgent query has been 'are we safer?'. This, however, is the wrong question. Of course, we are 'safer' – the posting of a single security guard at one building's entrance enhances safety, however microscopically. The correct question is 'are the gains in security worth the funds expended?'. Or, as it was posed shortly after 9/11 by risk analyst Howard Kunreuther, 'How much should we be willing to pay for a small reduction in probabilities that are already extremely low?'.[18] Working to answer this absolutely central question involves dealing with considerations of cost per saved life and acceptable risk as fed into cost–benefit methodology.

As far as we can see, Department of Homeland Security (DHS) decision-makers do not follow robust risk assessment methodology. If they did, low cost solutions that are easily deployed and effective would be the first to be implemented, and we do not find this to be the standard. This observation is supported by a committee of the US National Academy of Sciences in a 2010 report. After spending the better part of two years investigating the issue, the committee could not find 'any DHS risk analysis capabilities and methods' adequate for supporting the decisions made about spending on terrorism, and noted that 'little effective attention' was paid to 'fundamental' issues. With one exception, it was never shown 'any document' that could explain 'exactly how the risk analyses are conducted' and it looked over reports in which it was not clear 'what problem is being addressed'. This situation is particularly strange because, as the committee also notes, the risk models used in the department for *natural* hazards are 'near state of the art' and 'are based on extensive data, have been validated empirically, and appear well suited to near-term decision needs'.[19] Moreover, when it comes to terrorism, DHS appears to be exceptionally risk-averse: its decisions cannot be supported even with the most risk-averse utility functions possible, and its level of risk aversion is exhibited by few, if any, other government agencies, including the US Nuclear Regulatory Commission and the Environmental Protection Agency.[20]

For example, only once, as far as we can see, has DHS actually, if accidentally, engaged in a public assessment of acceptable risk. This involved the risk that body scanners that use X-ray technology will cause cancer. Asked about this, the DHS official in charge, John Pistole, essentially said that, although the cancer risk was not zero, it was acceptable. A set of studies, he pointed out, 'have all come back to say that the exposure is very, very minimal' and 'well, well within all the safety standards that have been set'.[21] Since the radiation exposure delivered to each passenger is known, one can calculate what the risk of getting cancer is for a single exposure using a standard approach that, though controversial, is officially accepted by nuclear regulators in the US and elsewhere. Based on a 2012 review of scanner safety, the fatal cancer risk per scan is about one in 60 million.[22] As it happens, the chance that an individual airline passenger will be killed by terrorists is much lower – one in 80 million.[23] Therefore, unless one believes that terrorists will in the near future become far more capable of downing airliners than they have been in the past, the risk of being killed by a terrorist in an airliner is already fully acceptable by the standards applied to the cancer risk from body scanners using X-ray technology. But no official, of course, has drawn that comparison.

Looking more broadly, any responsible analysis must also include a consideration of what else could have been done with the effort and money being expended on the policy proposed.[24] When we spend resources on regulations and procedures that save lives at a high cost, we forgo the opportunity to spend those same resources on measures that can save more lives at the same cost or even a lower one.[25] Homeland security expenditures invested in a wide range of more cost-effective risk reduction programmes like flood protection, vaccination and screening, vehicle and road safety, health care, and occupational health and safety would probably result in far more significant benefits to society. For example, diverting a few per cent of the nearly $10 billion per year spent on airline security could save many lives at a fraction of the cost if it were instead spent on seat belts for cars, bicycle helmets for children, tandem mass spectrometry screening programmes, front air bags, smoke alarms or tornado shelters.[26]

It may be useful in this light to put counterterrorism expenditures in the broadest comparative context. Björn Lomborg assembled a group of international experts to answer one question: 'if we had an extra $75 billion to put to good use, which problems would we solve first?'.[27] This sum is less than the amount the US spends on homeland security in a single year. More than 40 experts, tasked to do 'what is rational instead of what is fashionable', applied cost–benefit thinking to a wide range of issues. For many measures the benefit is ten times greater than the cost and, most importantly, the number of lives saved is spectacular. According to these analysts, an investment of merely $2 billion could save more than 1.5 million lives: one million child deaths could be averted by expanded immunisation coverage, while community-based nutrition programmes could save another half a million. In assessing expenditures for dealing with transnational terrorism, by contrast, the experts found costs to be three to 25 times higher than any benefits.[28]

If diversions of funds would easily save many lives, a government obliged to allocate funds in a manner that best benefits society must explain why it is spending billions of dollars on security measures with very little proven benefit and why that policy is something other than a reckless waste of resources. This disregard of basic cost–benefit considerations not only wastes money but costs lives.

We recognise that risk and cost–benefit considerations should not be the sole criterion for public decision-making. Nonetheless, they provide important insights into how security measures may (or may not) perform, their effect on risk reduction and their cost-effectiveness. They can reveal wasteful expenditures and allow limited funds to be directed to where the most benefit can be attained. If risk and cost–benefit advice is to be ignored, the onus is on public officials to explain why this is so and to detail the trade-offs and cuts to other programmes that will inevitably ensue.

'Policy-making is a risky business', one group of analysts has acknowledged. But they continue, 'regardless of the varied desires and political pressures, we believe that it is the responsibility of analysts forcefully to advocate rational decision methods in public policy-making, especially for those with high risk'.[29] Or, as Paté-Cornell observes, if rational approaches to public policy-making are not utilised, politically driven processes 'may lead to raising unnecessary fears, wasting scarce resources, or ignoring important problems'.[30] Central in all this, as risk analyst David Banks has suggested, is 'the distinction between realistic reactions to plausible threats and hyperbolic overreaction to improbable contingencies'.[31]

To be irrational with your own money may be to be foolhardy, to give in to guilty pleasure, or to wallow in caprice. But to be irrational with other people's money, particularly where public safety is concerned, is to be irresponsible. In the end it becomes a dereliction of duty that cannot be justified by political pressure, bureaucratic constraints or emotional drives. Risk reduction measures that produce little or no net benefit to society or produce it at a very high cost cannot be justified on rational life-safety and economic grounds. They are not only irresponsible but, essentially, immoral.

Finally, even if officials cannot bring themselves to embrace appropriate methodology to determine the value of their spending on security measures, they still bear a fundamental responsibility to inform the public honestly and accurately of the risk that terrorism presents. Instead, the emphasis has been on exacerbating fears. Indeed, despite the importance to responsible policy of seeking to communicate risk, and despite the costs of irresponsible fear-mongering, just about the *only* official in the US who has ever openly put the threat presented by terrorism in some sort of context is New York's Mayor Michael Bloomberg, who in 2007 pointed out that people should 'get a life' and that they have a greater chance of being hit by lightning than of being struck by terrorism.[32] It might be noted that this unconventional outburst did not have negative consequences for him. Although he had some difficulties in his re-election two years later, his blunt, and essentially accurate, comments about terrorism were not the cause.

Notes

1. See, for example, ISO 31000–2009, *Risk Management: Principles and Guidelines* (Geneva, 2009); and Peter L. Bernstein, *Against the Gods: The Remarkable Story of Risk* (New York: John Wiley, 1996).

2. M. Elisabeth Paté-Cornell, 'Quantitative Safety Goals for Risk Management of Industrial Facilities', *Structural Safety* 13 (1994): 145.
3. Lisa A. Robinson et al., 'Valuing the Risk of Death from Terrorist Attacks', *Journal of Homeland Security and Emergency Management* 7, no. 1 (2010) . Characteristically court awards or compensation payouts are considerably lower. The average life insurance payout to 9/11 victims was $350,000, and workers' compensation was $400,000. Lloyd Dixon and Rachel K. Stern, *Compensation for Losses from the 9/11 Attacks* (Santa Monica, CA: RAND Institute for Civil Justice, 2004), 31, 17. Court awards may be higher than this, but not always. A study of aviation fatality payments found that the average compensation for cases that went to trial was roughly $1.2 million, with a maximum of $10 million in 2010 dollars; however, half of all payouts, including those settled before trials began, were less than $350,000. James S. Kakalik et al., *Costs and Compensation Paid in Aviation Accident Litigation* (Santa Monica, CA: RAND Institute for Civil Justice, 1988), x. Payments to the families of soldiers killed in the Iraq war total $500,000, up from $112,240 before that conflict. Joseph E. Stiglitz and Linda J. Bilmes, *The Three Trillion Dollar War: The True Cost of the Iraq Conflict* (New York: W.W. Norton, 2008), 17.
4. A useful comparison might be made with the Los Angeles Police Department, which operates with a yearly budget of $1.2 billion. Mayor Antonio R. Villaraigosa, *Budget for the Fiscal Year 2013–14* (City of Los Angeles, April 2013), 152. Considering only lives saved following the discussion above, that expenditure would be justified if the police every year saved some 170 lives when each saved life is valued at $7 million. At present there are some 300 homicides each year in the city and about the same number of deaths in car accidents. It is certainly plausible to suggest that both these numbers would be substantially higher without police efforts and accordingly that local taxpayers are getting pretty good value for their money. Moreover, the police provide a great many other services to the community for the same expenditure, from directing traffic to arresting burglars and shoplifters. Although efforts to police terrorism also provide such co-benefits, they are likely to be quite a bit lower than those provided by the Los Angeles police.
5. C. C. Travis et al., 'Cancer Risk Management: A Review of 132 Federal Regulatory Decisions', *Environmental Science and Technology* 21, no. 5 (1987): 415–20.
6. See also Kenneth T. Bogen and Edwin D. Jones, 'Risks of Mortality and Morbidity from Worldwide Terrorism: 1968–2004′, *Risk Analysis* 26, no. 1 (2006): 56; and Daniel Gardner, *The Science of Fear: Why We Fear the Things We Shouldn't—and Put Ourselves in Greater Danger* (New York: Dutton, 2008), 250–1.
7. See, for example, National Consortium for the Study of Terrorism and Responses to Terrorism, *Integrated United States Security Database (IUSSD): Data on the Terrorist Attacks in the United States Homeland, 1970 to 2011*, http://www.start.umd.edu/start/publications/START_IUSSDDataTerroristAttacksUS_1970-2011.pdf (accessed December 11, 2013).
8. Michael Kenney, '"Dumb" yet Deadly: Local Knowledge and Poor Tradecraft among Islamist Militants in Britain and Spain', *Studies in Conflict & Terrorism* 33, no. 10 (2010): 911; and John Mueller and Mark G. Stewart, 'The Terrorism Delusion: America's Overwrought Response to September 11', *International Security* 37, no. 1 (2012): 81. For a discussion of the real-world relevance of *Four Lions*, a fictional film about a bumbling band of would-be terrorists in Britain, see John Mueller, 'Introduction', *Terrorism Since 9/11: The American Cases*, ed. John Mueller (Columbus: Mershon Center, Ohio State University, 2013), 27–8, http://polisci.osu.edu/faculty/jmueller/since.html.
9. See *US v. Faisal Shahzad* (2010, USDC, SDNY), http://s3.amazonaws.com/nytdocs/docs/333/333.pdf.
10. For a fuller discussion and for the derivation of this number, see John Mueller and Mark G. Stewart, *Terror, Security, and Money: Balancing the Risks, Benefits, and Costs of Homeland Security* (New York: Oxford University Press, 2011), chap. 4.
11. Ibid., chaps. 6–7; Mark G. Stewart and John Mueller, 'Aviation Security, Risk Assessment, and Risk Aversion for Public Decisionmaking ', *Journal of Policy Analysis and Management* 32, no. 3 (2013): 615–633; and Mark G. Stewart and John Mueller, 'Terrorism Risks and Cost–Benefit Analysis of Aviation Security', *Risk Analysis* 33, no. 5 (2013): 893–908.
12. Office of Management and Budget, 'Guidelines and Discount Rates for Benefit–Cost Analysis of Federal Programs (Revised)', Circular No. A-94, Washington, DC, 1992.

13. Mark G. Stewart and Robert E. Melchers, *Probabilistic Risk Assessment of Engineering Systems* (London: Chapman & Hall, 1997), 208–216; Paul Slovic, Baruch Fischhoff, and Sarah Lichtenstein, 'Facts and Fears: Understanding Perceived Risk', in *Societal Risk Assessment: How Safe is Safe Enough?*, eds. Richard C. Schwing and Walter A. Albers (New York: Plenum, 1980), 181–216.
14. James Fallows, 'If the TSA were Running New York', theatlantic.com, May 2010.
15. On this issue, see Mueller and Stewart, *Terror, Security, and Money*, 179–82. One might also compare the reaction to 9/11 with that to the worst terrorist event in the developed world before then, the downing of an Air India airliner departing Canada in 1985, in which 329 people, 280 of them Canadian citizens, perished. Journalist Gwynne Dyer points out that, proportionate to population, the losses were almost exactly the same in the two cases. But, continues Dyer, 'here's what Canada didn't do: it didn't send troops into India to "stamp out the roots of the terrorism" and it didn't declare a "global war on terror." Partly because it lacked the resources for that sort of adventure, of course, but also because it would have been stupid.' Gwynne Dyer, 'The International Terrorist Conspiracy', 2 June 2006, http://gwynnedyer.com.
16. Mueller and Stewart, *Terror, Security, and Money*, 87–92.
17. For a discussion, see John Mueller, Mark G. Stewart, and Benjamin H. Friedman, 'Finally Talking Terror Sensibly', 24 May 2013, http://nationalinterest.org.
18. Howard Kunreuther, 'Risk Analysis and Risk Management in an Uncertain World', *Risk Analysis* 22, no. 4 (2002): 662–3. See also John Mueller, 'Some Reflections on What, if Anything, "Are We Safer?" might Mean', 11 September 2006, http:// cato-unbound. org.
19. National Research Council of the National Academies, *Review of the Department of Homeland Security's Approach to Risk Analysis* (Washington, DC: National Academies Press, 2010). An evaluation of a risk analysis tool developed for the DHS is similarly critical. The tool has 'thousands of input variables', many of which cannot be estimated with much precision, and it could generate results that are 'completely wrong'. Moreover, it takes so long to run that it was not possible 'to conduct even a superficial sensitivity analysis' of its 'many thousands of assumptions and parameter estimates'. In addition, it only deals with relative risk, not absolute risk (a key criticism as well in the 2010 National Research Council study), and its estimates of these 'are subject to strong, probably untenable, assumptions'. The tool is also insensitive to changes in the magnitude of risk and 'assumes no attack can be deterred'. Andrew R. Morral et al., *Modeling Terrorism Risk to the Air Transportation System* (Santa Monica, CA: RAND Corporation, 2012).
20. Mark G. Stewart, Bruce R. Ellingwood, and John Mueller, 'Homeland Security: A Case Study in Risk Aversion for Public Decision-Making', *International Journal of Risk Assessment and Management* 15, nos. 5– 6 (2011): 367–86; and Stewart and Mueller, 'Aviation Security'.
21. PBS NewsHour, 16 November 2010, http://www.pbs.org/newshour/bb/transportation/july-dec10/airsecurity_11-16.html.
22. Scientific Committee on Emerging and Newly Identified Health Risks (SCENIHR), *Health Effects of Security Scanners for Passenger Screening* (Brussels: European Commission, 26 April 2012). Passenger exposure to backscatter scanners is 0.4 mSv per scan. A 1 mSv dose, according to standard models, increases the risk of fatal cancers by 0.004%. The increase in fatal cancer risk per scan is thus $0.4 \times 0.001 \times 0.004\% =$ one in 60 million.
23. Three hundred and sixty-three airline passengers were killed by terrorists in the 14 years between 1999 and 2012. The total number of global airline passengers in same period was roughly 29 billion.
24. Bruce Schneier, *Beyond Fear: Thinking Sensibly about Security in an Uncertain World* (New York: Copernicus, 2003).
25. Tammy O. Tengs and John D. Graham, 'The Opportunity Costs of Haphazard Social Investments', in *Life-saving, Risks, Costs, and Lives Saved: Getting Better Results from Regulation*, ed. Robert William Hahn (Washington, DC: American Enterprise Institute, 1996), 167–82.
26. Mueller and Stewart, *Terror, Security, and Money*, 182–3.
27. Björn Lomborg, *Global Crises, Global Solutions* (Cambridge: Cambridge University Press, 2009), 1.
28. Todd Sandler, Daniel G. Arce, and Walter Enders, 'Transnational Terrorism', in Lomborg, *Global Crises*, 552. They place the value of life at $2 million in their calculations.
29. J. Brian Hardaker, Euan Fleming, and Gudbrand Lien, 'How should Governments make Risky Policy Decisions?', *Australian Journal of Public Administration* 68, no. 3 (2009): 256–71.

30. Elisabeth Paté-Cornell, 'Risk and Uncertainty Analysis in Government Safety Decisions', *Risk Analysis* 22, no. 3 (2002): 633–46.
31. David L. Banks, 'Statistics for Homeland Defense', *Chance* 15, no. 1 (2002): 10.
32. Sewell Chan, 'Buzz over Mayor's "Get a Life" Remark', nytimes.com, June 6, 2007. See also John Mueller, *Overblown* (New York: Free Press, 2006), 151–2.

Homeland security: definitions and accountability

Amos N. Guiora

S. J. Quinney College of Law, University of Utah, USA

Counter-terrorism measures must be developed that are both effective and meet accountability standards. This article approaches the issue by: (1) proposing a matrix facilitating the measurement of the effectiveness or ineffectiveness of a particular counter-terrorism measure; and (2) proposing an accountability matrix for measuring the effectiveness of particular measures. The article's central focus is on developing criteria for measuring effective counter-terrorism – premised on the rule of law, policy considerations and the limits of power. If utilised, these criteria will provide empirical evidence that particular counter-terrorism measures in actuality contribute to effective and legal counter-terrorism.

Introduction

The terrorist attacks of September 11, 2001 left Americans feeling vulnerable as never before. For weeks thereafter citizens on the streets of the nation's capital glanced nervously skyward in response to noises from above. Even in seemingly safe communities across the heartland people thought twice before going out to shopping malls or restaurants. A year later Americans were still on edge. Comments by the Secretary of the Department of Homeland Security (DHS) created mass purchasing of duct tape and plastic sheeting, and Washington suburbanites brooded over whether the Washington sniper was the tip of al-Qaeda's latest plot. Arrests of shoe-bomber Richard Reid (2001), the Lackawanna Group (2002), terrorist cells in Toronto and Miami (June 2006), the UK-based group plotting to blow up US-bound airplanes (August 2006), the Times Square bomber (May 2010) and most recently the Boston Marathon bombers (April 2013) were reminders that the threat to the US homeland has not gone away.

In response to this threat the US government must determine how to implement effective counterterrorism measures. In order to do so, lawmakers must first determine criteria for measuring effectiveness primarily based on goals the policy seeks to achieve. If a government agency can empirically demonstrate that a particular security measure will contribute to effective and legal counterterrorism, then it has merit. However, if the measure is yet another example of groping in the dark or creating ad hoc counterterrorism measures not based on reliable and viable intelligence, then such measures will be ineffective and should be ruled non-starters. Lawmakers cannot wash their hands of the problem once they have established criteria to measure effectiveness. They must demand accountability from the executive; active legislative oversight is a critical component of this. Without stringent accountability requirements there is no guarantee that the executive will adhere to nor abide by the effectiveness criteria. This paper will address two core issues relevant to the broader homeland security discussion: framing homeland security; and articulating criteria for effectiveness and accountability in homeland security.

That said, the article focuses on the Bush administration's initial forays into homeland security; while not intended to serve as a historical document, it seeks to articulate the basis for the foundation of the DHS in the aftermath of September 11. Therefore the focus will be on the past; while, perhaps, shedding light on the present and the future, the primary purpose is a reflection on Bush administration policies. Furthermore, in large part, the Obama administration has not made overwhelming changes to the philosophical underpinnings of the DHS. While the challenges have changed, and are now increasingly focused on 'lone wolf actors',[1] the essence of DHS work remains largely similar between the two administrations. It is for this reason, then, that, in addressing homeland security, this article takes a look back.

Framing homeland security

The author proposed the following definition of homeland security when invited to testify before the US Congress in 2008:

> A group of preventative measures undertaken by a state in an attempt to reduce the probability that a terrorist attack will occur. This strategy will be fluid, constantly reassessing the balance between rights of the individual and rights of the state. A realistic strategy must prioritize threats according to their probability and imminence.[2]

For its part the US federal government has defined homeland security as 'a concerted national effort to prevent terrorist attacks within the United States, reduce America's vulnerability to terrorism, and minimize the damage and recover from attacks that do occur'.[3] The Department of Defense (DoD) clarified that:

> An external threat or aggression is an action, incident, or circumstance that originates from outside the boundaries of the homeland. Threats planned, prompted, promoted, caused, or executed by external actors may develop or take place inside the boundaries of the homeland. The reference to external threats does not limit where or how attacks may be planned and executed.[4]

The following is the Bush administration's rational for its homeland security strategy:

> The United States, through a concerted national effort that galvanizes the strengths and capabilities of Federal, State, local, and Tribal governments; the private and non-profit sectors; and regions, communities, and individual citizens—along with our partners in the international community—will work to achieve a secure homeland that sustains our way of life as a free, prosperous, and welcoming America. In order to realize this vision, the United States will use all instruments of national power and influence—diplomatic, information, military, economic, financial, intelligence, and law enforcement—to achieve our goals to prevent and disrupt terrorist attacks; protect the American people, critical infrastructure, and key resources; and respond to and recover from incidents that do occur. We also will continue to create, strengthen, and transform the principles, systems, structures, and institutions we need to secure our Nation over the long term. This is our strategy for Homeland Security.[5]

This definition of homeland security propagated by the federal government is unworkable and unrealistic for two major reasons: First, it does not clarify the need to consider the rights of the individual in relation to rights of the state. Second, and directly related to the current definition's lack of consideration for individual rights, the federal government must be realistic about its ability to prevent and control all threats. In this way priorities must be established according to the limits, both ideologically and fiscally, that the American people will support. In examining government policy in the aftermath of 9/11 the lack of a concentrated and realistic focus is dramatically apparent. In seeking to address all possible threats, the policy is, in actuality, not a policy. As documented below. the approach represents the antithesis of prioritisation, threat assessment and risk assessment. In seeking to protect against all threats at all times, the administration policy can best be described as overreaching, with all but inevitable violations of civil rights and liberties. The discussion below suggests a wish list rather than a systematic approach that critically examines prioritisation, threat assessment and risk assessment in the context of costs and benefits.

Therefore it is necessary to propose a sounder version of a homeland security strategy and policy fundamentally predicated on prioritisation, threat assessment and risk assessment in the context of costs and benefits that reflects balancing civil rights with the equally legitimate rights of the state. However, before engaging in that discussion it is important to set the stage by articulating the present state of (non)-policy.

Differentiating homeland security from other terms

An attempt to further refine the definition of 'homeland security' might best begin by distinguishing it from similar terms in the counterterrorism lexicon, specifically 'homeland defence' and 'counterterrorism'.

Homeland defence

As defined by the DoD in its *Strategy for Homeland Defense and Civil Support*, 'homeland defence' is the protection of US sovereignty, territory, domestic population and critical defence infrastructure against external threats and aggression, or other threats as described by the president.[6] The DoD clarified that external threats may arise internally, and the term 'external' does not limit where or how attacks might be executed.[7]

Counterterrorism

This comprises the actions of a state, proactive or reactive, intended to kill or injure terrorists and/or to cause serious significant damage to the terrorist's infrastructure.[8]

The difference between 'homeland defence' and 'homeland security' might correspond roughly to the difference between reaction and prevention. As Professor Wayne McCormack points out, 'while the political world and the media sometimes lump all responses to terrorism together under the label counterterrorism, that term is more appropriately used for prosecutorial and military-type responses while preventative measures are more aptly characterized as anti-terrorism.'[9] Homeland security, as defined by the federal government, meshes better with the concept of anti-terrorism, or prevention of terrorism.

Homeland security as distinct from national defence was established in Executive Order 13228, which created the Office of Homeland Security within the Executive Office of the President. This order identifies key homeland security functions: to detect, prepare for, prevent, protect against, respond to, and recover from terrorist attacks within the US.[10]

The federal government's definition of homeland security articulated above[11] brings into sharp focus a problem that will be tackled later in this project, namely, how can the nation make a concerted effort without treading too heavily on individual rights and liberties? But, for now, it might be best to think of homeland security as a group of preventative measures undertaken by a state in an attempt to reduce the probability that a terrorist attack will actually occur.

Structuring homeland security

Numerous state, federal and municipal agents work together to ensure public safety in the US. These include agents from law enforcement agencies, the military and intelligence gathering and analysis realms, public health, and emergency response sectors, who coordinate activities with the community's utilities, infrastructure, transportation, police and fire personnel. Some argue that these agencies should focus on less tangible issues, including job security and education, and that community values should be included in homeland security.

Executive branch documents under the Bush administration named two particular areas the US must be protected against: first, al-Qaeda, its affiliates and those inspired by them; and, second, catastrophic events, including natural disasters and human-made accidents.[12] Scholars have suggested instead three priorities with respect to homeland security: border security, critical infrastructure protection and intelligence analysis.[13]

Preventing and disrupting terrorist attacks

The Bush administration articulated as one of its most important tasks denying terrorists access to weapons of mass destruction (WMD) by detecting, disrupting and interdicting the movement of WMD-related materials. The administration stated: 'By integrating the operational and intelligence efforts of all levels of government, the private sector, and our foreign partners, and enabled by an international framework and domestic institutions supporting its implementation, our strategy involves simultaneous action to':

- determine terrorists' intentions, capabilities, and plans to develop or acquire WMD;
- deny terrorists access to the material, expertise, and other enabling capabilities required to develop WMD;
- deter terrorists from employing WMD;
- detect and disrupt terrorists' attempted movement of WMD-related materials, weapons, and personnel;
- prevent and respond to a WMD-related terrorist attack; and
- define the nature and source of a terrorist-employed WMD device.[14]

Intelligence plays a vital and imperative role in preventing terrorist attacks. Intelligence-led Policing (ILP) is based on community-oriented policing and problem solving, which the field of law enforcement has considered beneficial for many years.[15] The Foreign Intelligence Surveillance Act (FISA) provides a legal framework for the collection of foreign intelligence information.[16] The Protect America Act of 2007[17] allowed FISA to account for modern changes in technology and the threats from these.[18]

Border control and immigration

Control and management of the nation's borders and immigration into the nation are critical in achieving effective homeland security. The Bush administration decided to take the following acts to secure this area:[19]

- improve ability to detain and remove criminal and fugitive aliens and visa violators;
- hire, train and deploy additional Border Patrol agents, Customs and Border Protection officers and Immigration and Customs Enforcement officers;
- build on the substantial improvements to the infrastructure and technology deployed at our borders;
- expand detention bed space for aliens subject to detention and removal;
- enhance interior enforcement efforts, including worksite enforcement programmes.

The Bush administration also wanted employers to verify the work eligibility of all employees, preventing illegal immigrants from obtaining jobs through fraud or the use of stolen identification, including Social Security numbers. In order to accomplish this aim, the following measures were suggested:[20]

- expand the use of an electronic employment eligibility verification system that is timely, accurate and easy for employers to use;
- crack down on employers who knowingly hire illegal immigrants by applying criminal penalties to those who circumvent the law;
- step up efforts to verify the status of non-immigrants studying in the US through the Student and Exchange Visitor Information System (SEVIS) and have appropriate follow-up where there may be violations.

Some scholars believe that the Bush administration's border security plan is acceptable.[21] However, these same scholars have noted that non-homeland security missions, such as the Immigration and Naturalization Service's citizenship functions, should not be within the Homeland Security Department. Analysts have graded border and immigration activities as a C.[22]

Strengthen the foundation of homeland security

Several challenges are inhibiting the success of current homeland security efforts. Staff support for senior leaders is inadequate,[23] the Department has meagre means to manage the international dimensions of homeland security, and the Department is not staffed to reach out to other executive departments on an ongoing basis to handle foreign inquiries.[24] Of critical importance is the failure of a department-wide training system.

Reliance on 'on-the-job' training is not a sustainable approach to preparing front-line agents to simultaneously enforce immigration, customs and agriculture laws; to work with sophisticated technologies; and to be able to work in specialised assignments, such as being deployed to an overseas port as a part of the Container Security Initiative.[25]

The Department also has immature relations with the private sector regarding critical infrastructure protection. Currently, because security requirements have not been identified, private companies are not investing in comprehensive security measures to protect infrastructure they depend on, because their executives worry that such investments will place them at a competitive disadvantage.[26]

Critical infrastructure

The Office of Homeland Security has identified 17 sectors of critical infrastructure and key resources:[27]

1. agriculture and food;
2. banking and finance;
3. chemical (analysts have graded chemical plant security a D–F[28]);
4. commercial facilities;
5. commercial nuclear reactors, materials and waste;
6. dams (analysts have graded bridges, tunnels and other infrastructure a C[29]);
7. defence industrial base (analysts have graded air defence a B[30]);
8. drinking water and water treatment systems;
9. emergency services;
10. energy (analysts have graded nuclear plant security a B/B+[31]);
11. government facilities;
12. information technology;
13. national monuments and icons;
14. postal and shipping services;
15. public health and healthcare;
16. telecommunications;
17. transportation systems (analysts have graded airport security a C+[32]).

Some define critical infrastructure more widely as the 'systems and assets, whether physical or virtual, so vital to the United States that the incapacity or destruction of such systems and assets would have a debilitating impact on security, national economic security, or any combination of those matters'.[33] The National Infrastructure Protection Plan of 2009 gives a comprehensive risk management framework and provides a coordinated approach for federal, state, local, and private sector security partners.[34]

Identity management systems

To effectively protect America, security personnel must be trained and able to correctly identify individuals who pose a threat to the nation's security. Challenges in this arena include standardising secure identification, biometrics, and privacy and information security.[35] Thus the government has sought consensus on the proper balance between the development of information systems to address terrorist threats and the privacy protection of individual data and information.[36]

The 9/11 Commission requested that the government address biometric identifiers for screening across agencies by designing a comprehensive system to address common problems and common standards.[37] This includes setting standards for a host of documents, including birth certificates, passports and driving licences.[38] Congress passed the REAL ID Act of 2005 to establish federal standards for the issuance of driving licenses.[39] The government also implemented biometrics into passport issuance in 2006. E-passports 'employ a multi-layered approach' to protecting electronic data from skimming, eavesdropping and tracking.[40] With the increasing reliance on the internet for business, addressing evolving issues in cyber-security is also critically important.[41]

The Federal Financial Institutions Examination Council (FFIEC) guides financial security.[42] For banks offering internet-based financial services, the guidance describes enhanced authentication methods that regulators expect banks to use when authenticating the identity of customers using online products and services. In addition, financial institutions are expected to achieve compliance with the guidance.[43]

Port security

Analysts have graded port security a D +,[44] in part because the Coast Guard is inadequately stocked with ships and shore-based communication systems. Since post-9/11 patrolling requirements have increased, this lack of resources has made the need for recapitalisation more urgent.[45] A number of initiatives seeks to mainstream this effort.

1. The Container Security Initiative creates a system to pre-screen and evaluate maritime containers before they are shipped from foreign ports through automated targeting tools, ensuring that high-risk cargo is examined or scanned.[46] Delayed implementation of this Automated Commercial Environment, which would effectively manage and police the growing volume and velocity of goods that enter the US economy each day, has damaged security.[47]
2. The Customs–Trade Partnership Against Terrorism (C–TPAT) is a US Customs and Border Protection programme whereby businesses voluntarily undergo a security review and adopt enhanced security measures to expedite shipping.[48]
3. The Megaports Initiative allows the US to collaborate with foreign trade partners to enhance their ability to scan cargo for radiological materials at major international seaports.
4. The Secure Freight Initiative secures the global supply chain while keeping legitimate trade flowing.[49]

Disaster response

Some scholars believe the Federal Emergency Management Agency (FEMA) should have remained an independent agency, while the federal government's emergency preparedness and response programmes should have been folded into it.[50] The Bush administration created a National Response Plan to establish a comprehensive, national, all-hazards approach to domestic incident management across a spectrum of activities, including prevention, preparedness, response, and recovery.[51] Analysts have graded the disaster response capability a C-.[52]

Effectiveness and accountability in homeland security

Effectiveness

It is argued that effective counterterrorism causes the terrorist infrastructure to suffer serious damage – including damage to finances, intelligence, resources or personnel – thereby preventing a particular, planned attack from going forth and/or postponing or affecting plans for future attacks while minimising collateral damage, exercising fiscal responsibility and preserving civil liberties. This definition incorporates the following premises: (1) terrorism is not 100% preventable; (2) counterterrorism must have a short-term (tactical) as well as a long-term (strategic) component; and (3) counterterrorism must be conducted while balancing competing interests of human life, financial cost and civil liberty.

Terrorism is not 100% preventable

Security analysts are wont to frame recommended counterterrorism measures in an effectiveness paradigm that demands foolproof safeguards. However, it must be clearly stated that terrorism is not 100% preventable. Simply because a terrorist attack succeeds does not mean existing counterterrorism measures are ineffective. The inverse is also true: the absence of terrorist attacks does not necessarily indicate that existing counterterrorism measures are effective.

Counterterrorism must have a short-term as well as a long-term perspective

If a counterterrorism strategy only targets short-term threats, it will probably overlook other (long-term) real threats. It is important to note that terrorist organisations define effectiveness through the prism of long-term strategic considerations.[53] To understand the terrorist mind-set, it is necessary to appreciate the determination, resilience and single-mindedness with which terrorists work. Terrorists are willing to engage in a war of attrition with enormous personal hardship for the individual and his immediate family to achieve specific goals. Counterterrorism, both strategically and tactically, must be premised on this reality. Engaging in a never-ending cycle of violence is one means by which terrorist organisations signal to various audiences (the general public, followers and the relevant government) their commitment to the cause. From the geopolitical perspective of terrorists, pressure exerted by an attacked and affected public on the relevant government justifies continued attacks on innocent civilians.

How terrorist leaders view the effectiveness of their actions is critical. The following email, dated 3 October 2001, from Osama bin Laden to Mullah Omar highlights the dilemmas faced by the US after 9/11: 'If it [the US, Amos N. Guiora (ANG)] refrains from responding to jihad operations, its prestige will collapse, thus forcing it to withdraw its troops abroad and restrict itself to US internal affairs. This will transform it from a major power to a third-rate power, similar to Russia.'[54]

Counterterrorism must be conducted in balance with competing interests of human life, financial cost and civil liberty

Finding a balance between national security and the rights of individuals is the most significant issue faced by liberal democratic nations developing a counterterrorism strategy. Without a balance between these two tensions, democratic societies lose the very ethos for which they fight. As Benjamin Franklin once said, 'those who would give up essential liberty, to purchase a little temporary safety, deserve neither liberty nor safety.'[55] Indeed, it is imperative for democracies to avoid infringing political freedoms and civil liberties. Yet a government's ultimate responsibility is protecting its citizens. This struggle to balance competing interests may be the most fundamental dilemma confronting democracies today.

The first step in creating an effective counterterrorism measure is analysing the threat. To that end, the following questions must be answered:

1. What is the threat the state faces?
2. Who is responsible for planning the threat?
3. Who is responsible for financing the threat?
4. Who is responsible for carrying out the threat?
5. When is the threat likely to be carried out?

Once these questions are answered, the threat can be placed on an imminence continuum with the understanding that one large threat may be comprised of smaller, more manageable, threats. The imminence continuum has four major benchmarks: imminent, foreseeable, long-range and uncertain. 'Imminent threats' are those that are to be shortly conducted; 'foreseeable threats' are those that will be carried out in the near future (with no specificity); 'long-range threats' are threats that may reach fruition at an unknown time; 'uncertain threats' constitute those that invoke general fears of insecurity.

Where a particular threat fits on the imminence continuum affects the balance that must be struck between national security and competing rights and interests. Figure 1 depicts the imminence continuum graphically. The threat of a terrorist attack is listed from left to right, progressing from imminent to uncertain. The vertical column on the left lists seven factors that counterterrorism measures must balance in considering these threats. The balancing factors include collateral damage, civil liberties, valid intelligence, frequency of reporting,

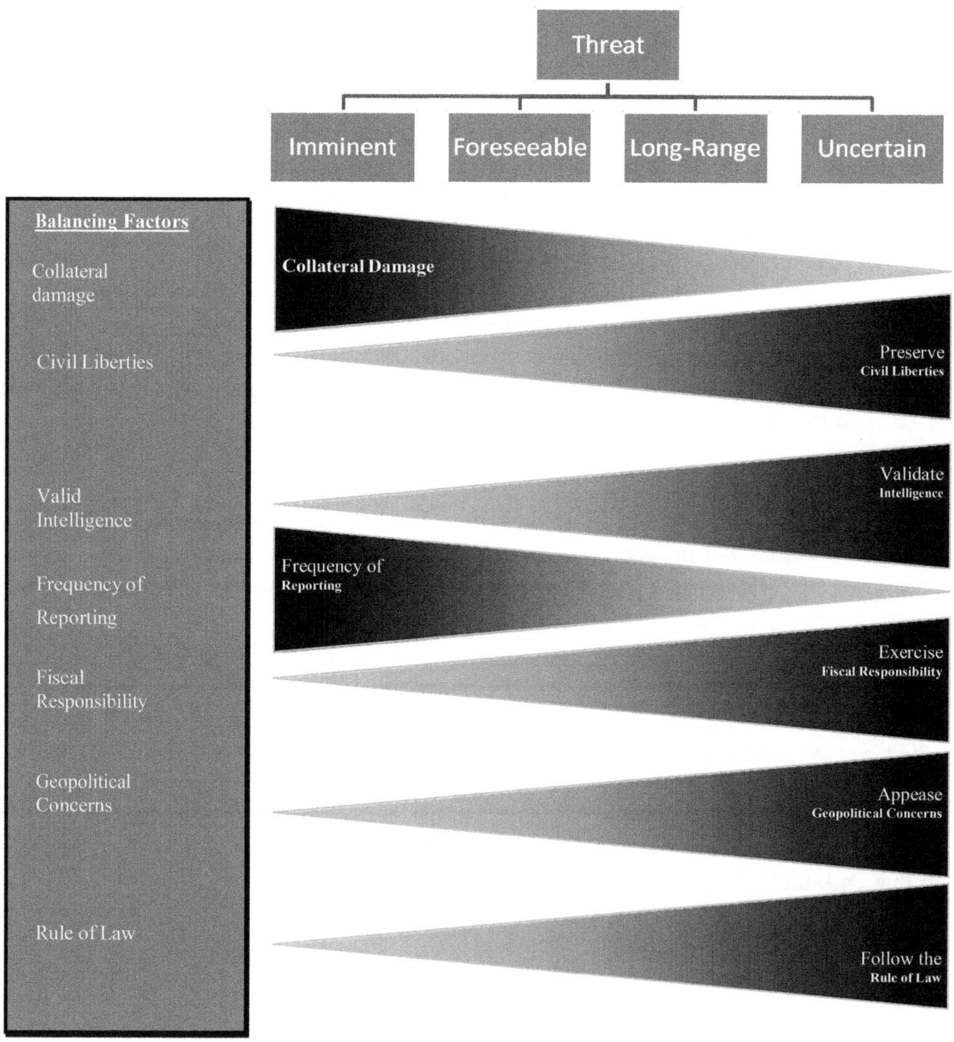

Figure 1. Threat and imminence.

fiscal responsibility, geopolitical concerns and the rule of law. Understanding these factors is crucial; detailed explanations are outlined below the chart. The triangular bars in the body of the graph represent the relative priority placed on each of these factors in the event of an imminent, foreseeable, long-range, or uncertain threat of a terrorist attack. The thicker the triangular bar, the greater the importance of the corresponding factor. For example, the triangular bar representing the first factor, collateral damage, is thicker for an imminent threat and thinner as it reaches an uncertain threat. This bar indicates that collateral damage is more likely for imminent counterterrorism measures than for foreseeable, long-range or uncertain measures..

Collateral damage

Collateral damage requires minimising loss of civilian life in a military operation; proportionality requires that civilian losses be proportional to the military advantage, which will be assessed as follows: what were the factors in target selection and what were the means and methods of attack?[56] Ultimately the public is willing to stomach greater collateral damage the more imminent a threat is.[57] However, alternative effective measures that would lessen collateral damage must also be considered. In doing so, collateral damage becomes only one of many factors to be weighed when selecting one measure over others.

Civil liberties

Liberal democratic societies that unilaterally decide on self-imposed restraints inherently limit their responses to terrorism.[58] However, balancing legitimate national security needs against the rights of those individuals living in the nation is a true test of a nation's adherence to democratic values. Possible suspension of constitutionally guaranteed liberties and rights must be weighed against legitimate national security considerations. That is a balancing dilemma that decision-makers must address.

Validating intelligence

To be valid, intelligence must be reliable, viable and corroborated.[59] Needless to say, reliable, viable and corroborated intelligence may be difficult to obtain. In all circumstances reasonable effort must be made to obtain valid intelligence before action is taken. However, the level of imminence dictates the definition of 'reasonable' (both with respect to the reliability of the information and to how much time is allotted to its corroboration).

Frequency of reporting

This factor encompasses the forthcoming accountability discussion. Congress has mandated annual reports on terrorist threats.[60] These reports, however, are too infrequent for gauging imminent and foreseeable threats and possibly inapplicable to uncertain threats. Counterterrorism measures taken to address imminent threats should be reported to Congress after the threat has passed or been addressed. Only in this way can Congress have meaningful review of significant counterterrorism measures. Foreseeable threats should be reported to Congress when the threat is identified. This reporting enables Congress to emphasise to the executive the need for balancing, and to better understand how different agencies are detecting and countering various threats. Congress has appropriately addressed the need for annual reporting regarding long-range threats. Conversely, many uncertain threats will not fall within the scope of the federal government.[61] Threats that do, however, may appropriately be addressed when agencies address long-range threats. State legislatures should implement reporting procedures for uncertain threats that exist on a state-level.

Fiscal responsibility

Financial costs necessarily limit the quantity of counterterrorism measures a nation may conduct. With limited resources government must pick and choose which measures will most effectively counter short-term and long-term threats. Although the need for fiscal responsibility lessens as a threat becomes more imminent, careful planning for potential attacks will allow careful use of financial resources even in the face of imminent threats. This would, thereby, free resources for countering long-range and uncertain threats. Thus, financial responsibility needs to be considered not only in light of the level of a particular threat, but also in light of an overarching counterterrorism strategy.

Geopolitical concerns

Counterterrorism necessarily raises international concerns because the threat does not necessarily reside exclusively within a nation's borders. It is recommended that the court of international opinion be considered when selecting a particular counterterrorism measure. Wrong choices cause the US to lose global influence or serve as an impetus that turns the swayable population into terrorists. If measures are chosen in concert with geopolitical concerns, then weight must be given to those that have the potential to make an impact on geopolitical considerations. Of course, in the light of an imminent threat, geopolitical concerns do not weigh the same as when establishing predicated deterrence strategies for uncertain threats, or strategies to counter long-range threats.

Rule of law

The rule of law protects free democracies and sets a basis for trust between nations. In order to best adhere to the rule of law, counterterrorism measures should be drafted in advance of such an actual threat. Such measures should dictate which laws may be relaxed, and to what extent, when facing an imminent threat. To wait until a threat is present denies government the opportunity to make careful and cautious choices that will provide security to the public, while balancing civil rights with legitimate national security considerations.

In summary, to conduct counterterrorism effectively, measures undertaken by the state must target both short-term and long-term terrorist operations. They must also balance the competing interests of national security with (1) minimal collateral damage; (2) civil liberties; (3) reliable, viable and corroborated intelligence; (4) fiscal responsibility; (5) frequent reporting to Congress; (6) geopolitical concerns; and (7) the rule of law. This balancing must occur in the context of the perceived threat's imminence.

Accountability

Accountability means articulating in a transparent manner the effectiveness or ineffectiveness of a particular counterterrorism measure or strategy to one's superiors with the power to rectify or discontinue measures. The 9/11 Commission Report emphasises in detail the need for standards of accountability in developing and implementing counterterrorism measures. The 9/11 Commission correctly stated that 'effective public policies … need concrete objectives'.[62] That is, in the struggle against terrorism, agencies need to be able to measure success:

> These measurements do not need to be quantitative: government cannot measure success in the ways that private firms can. But the targets should be specific enough so that reasonable observers – in the White House, the Congress, the media, or the general public – can judge whether or not the objectives have been attained. [. . .] Vague goals match an amorphous picture of the enemy. Al-Qaeda and its affiliates are popularly described as being all over the world, adaptable, resilient, needing little higher-level organisation, and capable of anything. The American people are thus given the picture of an omnipotent, unslayable hydra of destruction. This image lowers expectations for government effectiveness....We do not believe it is possible to defeat all terrorist attacks against Americans, every time and everywhere . . . But the American people are entitled to expect their government to do its very best. They should expect that officials will have realistic objectives, clear guidance, and effective organization. They are entitled to see some standards for performance so they can judge, with the help of their elected representatives, whether the objectives are being met. [63]

The 9/11 Commission Report appropriately called for these objective standards of accountability through which we can evaluate the effectiveness of the counterterrorism measures.

Without standards for accountability Congress unwittingly creates an unfettered executive. An unfettered executive, unrestrained by courts and legislatures, is detrimental to liberal democracies attempting to balance national security and individual rights.[64] Furthermore, when neither the legislature nor the judiciary rein the executive in, the former is bound to make mistakes, and therefore, more effective alternative means are often overlooked. Particularly in the murkiness and uncertainty of drawn-out amorphous operational counter-terrorism, the executive must know there are clear guidelines determining accountability. Therefore, in the fight against terrorism, maintaining a strict separation of powers, and checks and balances in government, is of supreme importance.

In the immediate aftermath of 9/11 Congress failed in its role of oversight and enabled an unfettered executive. Congress overwhelmingly ratified the Authorization for Use of Military Force (AUMF),[65] while failing to articulate limits of what and where the force could be used. Congress acted, but its actions were more akin to support than to oversight. Rather than engage the administration in debate regarding the limits of power, Congress stood by while the unfettered executive defined the debate. Congress must act proactively in order to limit executive power.

Matrix for implementing accountability

On a practical level Congress must ask questions that enable it to determine whether a particular counterterrorism measure should proceed as defined, should be altered, or should be discontinued. The ultimate conclusion (whether to continue, alter or discontinue a measure) cannot be reached by the answer to only one question, such as what is the financial cost? Rather, the conclusion will be reached by recognising a balance between numerous factors. For example, the nation may be willing to pay a higher financial price tag if the measure preserves civil liberties to a greater extent than less expensive alternatives.

The matrix below (Figure 2) lists the questions that must be answered before Congress decides whether a measure should be continued, altered or discontinued. The answer to each question is placed on a continuum, guiding Congress to its ultimate decision. After every question has been answered, Congress can view the matrix and determine if the measure strikes a balance between factors, should be altered to strike a better balance

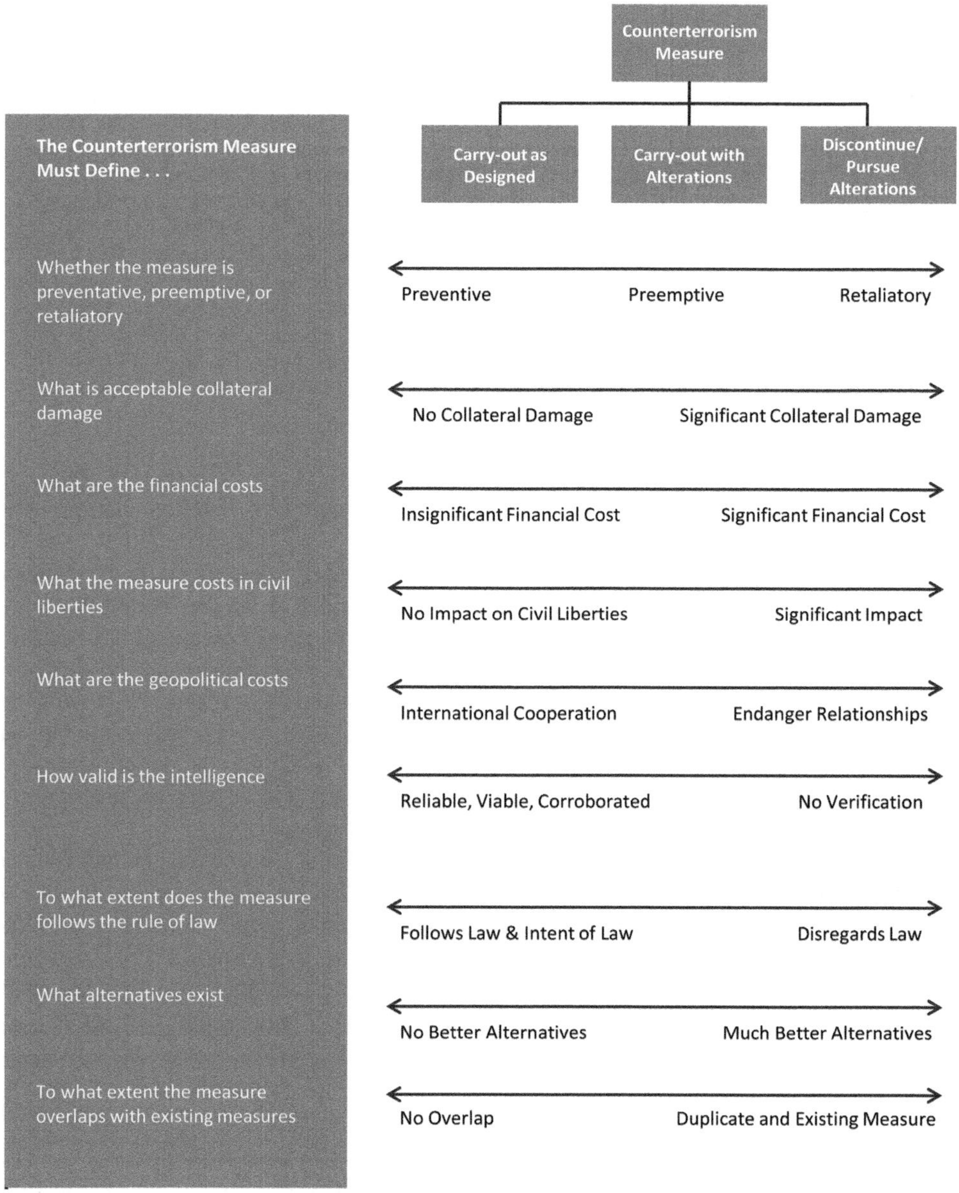

Figure 2. Matrix for implementing accountability.

based on where it is out of balance, or should be discontinued. These questions in the matrix, if asked in a timely manner, will lead to effective congressional oversight.

The matrix represents the three options available – continuance, alteration or discontinuance – graphically at the top of the chart, from left to right. On the left are the questions that Congress must ask to determine whether a measure is effective: (1) is the measure preventative, pre-emptive, or retaliatory? (2) what is acceptable collateral damage? (3) what are the financial costs? (4) what are the costs to civil liberties? (5) what are the geopolitical costs? (6) how valid is the intelligence? (7) to what extent does the measure follow the rule of law? (8) what alternatives exist? and (9) to what extent does the measure overlap with existing measures? The body of the matrix includes horizontal spectrums that allow for real-world answers to the accountability questions. Only after answering all the accountability questions can Congress determine whether the particular measure should be continued, altered or discontinued. For complicated measures that elicit mixed results, Congress can view the matrix and determine if the measure strikes a balance between answers that encourage implementation and answers that encourage discontinuance or non-implementation. The questions in the matrix, if asked in a timely manner, will lead to effective congressional oversight.

Is the measure preventative, pre-emptive, or retaliatory? Preventative measures counter terrorism before terrorists are prepared to strike. These measures work to prevent the swayable population (those not convinced terrorism is the answer) from becoming terrorists; destroy training camps and other terrorist infrastructure; and strengthen nations vulnerable to terrorists. Pre-emptive measures prevent particular terrorist acts from being carried out. Retaliatory measures are emotional responses to terrorist attacks. Preventative and pre-emptive measures can be considered self-defence, however; retaliatory measures are unlawful according to international law.[66]

What potential collateral damage does the measure cause? While the goal is to minimise collateral damage, this factor is weighed in light of the other factors. It is necessary to balance collateral damage with financial cost, cost to civil liberties and risk to personnel.

What is the financial cost of conducting the measure? Operational counterterrorism can be a financially costly endeavour. However, financial cost must be balanced with the effectiveness of the measure (including costs of civil liberties, collateral damage, the nature of the target, and more).

What is the cost to civil liberties of conducting the measure? Civil liberties define our democracy. When a measure proposes to suspend civil liberties, Congress must decide whether there are alternative measures that have a smaller impact on affected population groups. Unless circumstances, such as an imminent threat, warrant otherwise, counterterrorism measures that infringe civil liberties must be invalid. However, measures that may infringe (even minimally) civil liberties in an emergency context must be subject to strict legislative oversight and judicial review.

How valid is the intelligence that led to implementing the measure? To be valid, intelligence must be reliable (to what degree is the source reliable?), viable (to what degree can the threat actually be carried out?), and corroborated (who or what else provides similar intelligence?).

What geopolitical concerns arise as a result of the measure? Ideally the US would work in concert with the other nations of the world when conducting global operational counterterrorism. However, achieving global consensus and support can be time-consuming, costly and at times impossible. Yet the US must understand the risks it takes when it conducts a particular measure unilaterally. Decision-makers must determine if they are willing to accept the geopolitical consequences.

Does the measure follow the rule of law? To be valid, counterterrorism measures must be rationally based on constitutional and legal foundations. Congress has a duty to create laws that empower agencies but also to enact clear guidelines that limit the agencies' power. Clear guidelines will enable the judiciary to hold agencies accountable for their counterterrorism efforts. Further, clear guidelines will instil within Congress greater control over counterterrorism efforts.

What alternatives exist? Alternatives must always be identified and viable alternatives explored. Without recognising the alternatives, Congress cannot know whether a particular measure is actually cost-effective – financially, socially or politically.

To what extent does the measure overlap with existing measures? Although a measure may pass the other nine explorations without pause, this question is still significant. If a particular measure overlaps with another existing measure, then one of the two measures may be scaled back to the extent the two overlap to avoid unnecessary social and financial costs.

US government agencies and departments: evaluation of effectiveness and accountability

In the past the US has sought to conduct effective (though largely undefined) counterterrorism. However, the country's lack of a cohesive definition of effectiveness that strikes a balance between competing interests has largely produced scattered approaches, measures that have duplicative purposes, and gaps in security or intelligence.

Although various agencies are analysed below in turn, the same general analysis applies to all. The result of limited accountability to which the agencies are subject is a lack of clarity and focus regarding an agency's approach to counterterrorism. By requiring agencies to analyse identified threats and by discussing their efforts to balance achieving security with the other factors indicated above, this lack of clarity will be minimised, if not eliminated.

Department of Homeland Security. The DHS, focusing inside the borders of the US, works within the civilian sphere to protect the territory of the country from terrorist attacks and responds to natural disasters affecting the nation. According to Section 101 of the 2002 Homeland Security Act,[67] the principal mission of the Department is to: prevent terrorist attacks within the US; reduce the vulnerability of the US to terrorism; and minimise the damage, and assist in the recovery, from terrorist attacks that do occur within the country. To this end the Department's primary responsibilities include: (1) information analysis and infrastructure protection; (2) chemical, biological, radiological, nuclear and related countermeasures; (3) border and transportation security; (4) emergency preparedness and response; and (5) coordination (including the provision of training and equipment) with other executive agencies, with state and local government personnel, agencies and authorities, with the private sector, and with other entities.

DHS performance is analysed by the Office of Management and Budget's (OMB) Program Assessment Rating Tool (PART).[68] According to PART, 42% of surveyed DHS programmes were rated as either effective (14%) or moderately effective (28%); 25% were rated only adequate; the remainder, nearly 33%, yielded a 'results not demonstrated' classification.[69] Using a similar PART rating system, OMB determined that 13% of DHS spending (representing $5.4 billion) was effective, 53% ($22.2 billion) was moderately effective and 19% ($8 billion) was merely adequate; 15% ($6.2 billion) was counted as 'results not demonstrated'.[70]

While it is indisputable that progress has been made within the DHS, public criticism often contradicts the optimistic finding of the PART surveys, indicating that much remains to be done before all DHS programmes can be considered effective. Lack of organisation, resources and leadership contributed to DHS ranking last in job satisfaction among 36 surveyed government agencies in 2012. Homeland security experts claim the department's emphasis on law enforcement and security rules, as opposed to greater focus on intelligence and nuclear non-proliferation, are ultimately undermining the nation's security from terrorist threats. Many experts cite the failure of Congress to consolidate its oversight of homeland security (ignoring the 9/11 Commission's recommendation to do so) as a major impediment to effective action. Indeed, Secretary Chertoff has often cited the difficulty of protecting the nation while '[The DHS is] serving so many masters with inconsistent positions'.[71]

Federal Bureau of Investigation (FBI). The FBI counterterrorism programme is designed to protect the US from terrorist attacks. FBI agents, analysts and other staff investigate threats to prevent attacks from occurring, pursue sanctions against terrorists, and respond to terror attacks against Americans.[72] These purposes and objectives are very broad. The likelihood is therefore high that the FBI may duplicate the counterterrorism efforts of other agencies.

In the past the FBI's counterterrorism measures have been held accountable internally and by PART; various critics have also commented on the FBI's effectiveness. The OMB and other federal agencies state that the FBI counterterrorism programme is performing adequately, meaning that it needs to set more ambitious goals, achieve better results, improve accountability or strengthen its management practices.[73] The OMB states that the FBI has greatly expanded its counterterrorism programme, but only set performance measures two years ago. In those two years it fell short of its goals in several areas, including validating information provided by human sources and ensuring that sources are reporting on the most important subjects.

However, in 2004, FBI Director Robert S. Mueller, III, testified before the Select Committee on Intelligence of the United States Senate on 'Success in the War on Terrorism in 2003'. He highlighted the following as successes: locating and charging al-Qaeda operatives in Cincinnati and Baltimore; arresting the US leader of Palestine Islamic Jihad and his lieutenants for participating in a conspiracy that contributed to the deaths of two US citizens in Israel; indicting three people in Newark for at attempting to smuggle an SA-18 shoulder-fire missile system into the US; arresting and charging an individual in Minneapolis with providing funds to al-Qaeda in Pakistan; and interviewing more than 10,000 Iraqi expatriates to gain information in support of Operation Iraqi Freedom.[74]

The federal government and other critics agree that the FBI is facing agency-wide challenges. For example, the FBI does not tie its budget requests to its performance goals, and its audits show financial management system deficiencies that compromise the Bureau's ability to manage its programmes.[75] In the FBI's domestic counterterrorism work the primary product is information and analysis rather than the historical arrests and convictions, making it far more difficult to measure the productivity of agents. For FBI agents with many years of experience doing criminal work, the shift to counterterrorism has required a fundamental reorientation of daily activities and of how the agents judged their own performance.[76] Richard Posner and others argue that the agency-wide changes in the FBI have not gone far enough, leaving the FBI with a bias in favour of criminal work and a staff that prefers criminal work to counterterrorism investigations.[77]

To improve the FBI counterterrorism program, the government promises to ensure sources provide reliable information, increase training for counterterrorism personnel, and strengthen the links between budget requests and performance levels.[78]

Transportation Security Administration (TSA). The purpose of the Transportation Security Support of the TSA is to support the protection of the nation's transportation system by assessing and distributing intelligence information; providing comprehensive technical infrastructure support; and providing support to accomplish the overall TSA mission.[79] TSA has a fairly narrow counterterrorism mission. For this reason, it is imperative that the TSA communicate frequently with the broader intelligence gathering agencies to ensure that it is identifying all imminent, foreseeable, and long-range threats. Balancing civil rights and financial costs are the most predominant factors for TSA to consider.

As analysed by the OMB and other federal agencies, this programme has not yet demonstrated results because it has not developed acceptable performance goals or collected data on its performance.[80]

The government states that the following actions are underway to improve the program: establishing long-term goals for performance in Financial Year 2007 and evaluation in 2008; improving budgetary programme planning; and encouraging independent evaluations of its effectiveness.[81] This is consistent with the statement of Gerald L. Dillingham, Director of Civil Aviation Issues for the US General Accounting Office (GAO), in 2003 to the National Commission on Terrorist Attacks Upon the United States. He said that, although the TSA has made progress in addressing certain cost-related issues, it has not established an infrastructure that provides effective controls to monitor contractors' costs and performance. To ensure control over TSA contracts, The Department of Transportation's Inspector General has recommended that Congress set aside a specific amount of TSA's contracting budget for overseeing contractors' performance with respect to cost, schedule and quality.[82]

Central Intelligence Agency (CIA). The purpose of the CIA is to collect, analyse, evaluate and disseminate foreign intelligence to assist the president and senior US government policy-makers in making decisions relating to national security.[83]

The CIA employed post-911 operations to respond to terrorist-related threats, including rendition, detention, and interrogation programmes, as well as close collaboration with allied intelligence services. As of September 2007, 'fewer than 100 people had been detained at CIA's facilities ... the number of renditions ... is even a smaller number.'[84] The programmes are selective; they 'were designed only for the most dangerous terrorists and those believed to have the most valuable information, such as knowledge of planned attacks'.[85] Nevertheless, this activity remains highly controversial, especially outside the US.[86]

The Senate Select Committee on Intelligence and the House Permanent Select Committee on Intelligence requested that the CIA's Office of Inspector General (OIG) review the findings of the Joint Inquiry (JI) Report. The Accountability Review Team created a Report on Accountability of the CIA with respect to the 9/11 attacks.[87] The Accountability Review Team found neither a 'single point of failure' nor a 'silver bullet' that would have enabled the intelligence community to predict or prevent the 9/11 attacks. The team did find, however, failures to implement and manage important processes to follow through with operations, and to properly share and analyse critical data. While this report specifically deals with the CIA and al-Qaeda, it provides some important themes for criteria to examine the effectiveness of counterterrorism on a broader spectrum: create one comprehensive strategic plan to target a specific security threat; use all authorities to lead the effort against the specific security threat; ensure that the director receives regular updates on efforts to track and disrupt the specific security threat; ensure cooperation among all intelligence gathering agencies toward a common goal; and ensure that technology is used to full capacity to support effective intelligence gathering.

Clearly, the CIA recognises the need for valid intelligence. However, it must also be careful to balance civil rights, geopolitical concerns and financial costs. Such balancing will occur naturally if the agency identifies particular threats and labels them 'imminent', 'foreseeable', 'long-range', or 'uncertain'. Of course, congressional oversight as well as coordinating with other agencies will help to ensure a proper balance as well as eliminate duplicative measures.

Department of Defense (DoD). The DoD is charged with the protection of the people, assets, property and government of the US and is part of the Executive of Congress. As of 1986 legislation reorganised the DoD by enacting the Goldwater Nichols Act.[88] The Joint Chiefs of Staffs in charge of the various branches of the military answer to the Executive Branch, through the Secretary of Defense. The Executive Branch is constrained by the war powers of Congress, requiring an Act of Congress in order for the US to declare war.[89] In rare instances, and in light of serious domestic threats in a post-9/11 world, the DoD may take action domestically. The 'Secretary of Defense may ... make available any equipment (including associated supplies or spare parts), base facility, or research facility of the Department of Defense to any Federal, State, or local civilian law enforcement official for law enforcement purposes'.[90]

Measuring the DoD's effectiveness requires analysis of how quickly it accomplishes directions of the executive and Acts required by Congress to the satisfaction of the executive and Congress, and whether it achieves an optimal level of protection for the people, assets, property and government entities of the US.[91] An optimal level of protection requires computing cost of protection weighed against benefits.

Similarly to the FBI, the DoD has a broad mission. Because of the Department's strong visual presence, geopolitical concerns should be weighed more so than in other agencies. However, all balancing factors are relevant to the DoD.

Department of State. The Department of State, primarily via the Office of the Coordinator for Counterterrorism, has the lead role on the diplomatic front regarding counterterrorism. The Department works closely with other agencies and organisations to shut down terrorist financial networks, provide humanitarian aid, and to investigate terrorist organisations and activities and bring terrorists to justice.

Until recently the State Department's Office of the Coordinator for Counterterrorism has been criticised as a marginal player in the global war on terror.[92] The reasons have included lack of effective leadership, insufficient resources, and the difficulty of helping to coordinate the wide scope of counterterrorism efforts, all of which in turn engendered low morale and employee recruitment or retention difficulties. However, the most recent OMB PART surveys report that primary State Department counterterrorism efforts are, for the most part, performing effectively.

The Department's Terrorist Interdiction Program (international border control databases), Anti-terrorism Assistance Program (building international anti-terrorism relationships and providing counterterrorism assistance to key countries), Worldwide Security Upgrades Program (protecting American lives, property and information overseas and domestically), and the South Asia Military Assistance Program (providing military assistance to promote regional security) have all received effective ratings indicating that they have set ambitious goals and achieved results, are well-managed, and improve efficiency. The glaring exception is the Security Assistance for Near East Asia Program (promoting US national security by preventing and responding to terrorism by strengthening military-to-military ties in the region) which received an 'adequate' rating,[93] indicating that the programme needs to set more ambitious goals, achieve better results, improve accountability or strengthen its management practices.

Conclusion

The ultimate question is 'What price is this democracy willing to pay for security?'[94] Congress must answer this question. Indeed, Congress has the responsibility of holding the executive accountable for counterterrorism measures. In so doing, it must make precise and concise decisions. By giving unchecked power to the executive, Congress fails in its role to ensure that the US government is conducting counterterrorism effectively – a role bestowed upon it by the Constitution. Measuring effectiveness will be significantly facilitated by implementing the matrix proposed in Figuire 2. Furthermore, Congress needs not only to define how agencies determine whether their measures are effective, but must hold agencies accountable for their choices based upon the matrix model this paper recommends.

Acknowledgement

With thanks to Jason Shelton (JD expected 2014), S. J. Quinney College of Law, University of Utah, for his editorial assistance.

Notes

1. See, further, the article by Appleton in this issue.
2. Amos N. Guiora, 'Written Testimony Before a Hearing of the Committee on Homeland Security, US House of Representatives', in *The Resilient Homeland: How DHS Intelligence Should Empower America to Prepare for, Prevent, and Withstand Terrorist Attacks* (U.S. Government Printing Office, May 2008) 11; http://www.gpo.gov/fdsys/pkg/CHRG-110hhrg43940/html/CHRG-110hhrg43940.htm.
3. Homeland Security Council, *National Strategy for Homeland Security* (Washington, DC: HSC, 2007), 3, http://www.dhs.gov/xlibrary/assets/nat_strat_homelandsecurity_2007.pdf.
4. Joint Publication 3–27 , *Homeland Defense* (Washington, DC, 2013), I-1, http://www.dtic.mil/doctrine/new_pubs/jp3_27.pdf.
5. Homeland Security Council, *National Strategy for Homeland Security*, 13.
6. Department of Defense, *Strategy for Homeland Defense and Civil Support* (Washington, DC: DoD, 2005), 5, http://www.defenselink.mil/news/Jun2005/d20050630homeland.pdf.
7. Ibid.
8. Amos N. Guiora, *Global Perspectives on Counterterrorism* (New York: Aspen Publishers, 2007), 139.
9. Wayne McCormack, *Legal Responses to Terrorism* (San Francisco: LexisNexis/Matthew Bender, 2005), 10.
10. Executive Order No. 13228, 66 F.R. 51812, http://www.dojgov.net/13228.pdfx (accessed December 27, 2013).
11. Homeland Security Council, *National Strategy for Homeland Security*, 3. See also Department of Homeland Security, 'Strategic Plan Fiscal Years 2012–2016', 2 , Feb. 2012, https://www.dhs.gov/xlibrary/assets/dhs-strategic-plan-fy-2012-2016.pdf.
12. Homeland Security Council, *National Strategy for Homeland Security*, 1.
13. Paul Light and James Lindsay, Council on Foreign Relations, 'Views of Homeland Security', *Washington Times*, July 25, 2002, http://www.cfr.org/world/views-homeland-security/p6395.
14. Homeland Security Council, *National Strategy for Homeland Security*, 15.
15. Ibid., 19.
16. Foreign Intelligence Surveillance Act of 1978, 50 USC 1801, http://www.gpo.gov/fdsys/pkg/STATUTE-92/pdf/STATUTE-92-Pg1783.pdf.
17. Protect America Act of 2007, 50 USC 1801, http://www.intelligence.senate.gov/laws/pl11055.pdf.
18. Homeland Security Council, *National Strategy for Homeland Security*, 21.
19. Ibid., 16.
20. Ibid.
21. Light and Lindsay, 'Views of Homeland Security'.

22. Stephen Flynn, Council on Foreign Relations, 'Report Card', October 25, 2006, http://www.cfr.org/publication/11814/flynn.html. Stephen Flynn, Council on Foreign Relation's senior fellow for national security studies, considers the state of homeland security five years after the creation of the Office of Homeland Security, handing out grades A – F representing performance levels in several vulnerable key areas of homeland security. See also US Department of Homeland Security, 'Annual Performance Report for Fiscal Years 2010–2012', http://www.dhs.gov/xlibrary/assets/cfo_apr_fy2010.pdf; and Stephen Flynn, 'Recalibrating Homeland Security: Mobilizing American Society to Prepare for Disaster', *Foreign Affairs* 90, no. 3 (2011), 130–140. 'The Bush administration's emphasis on combating terrorism overseas meant that it devoted limited strategic attention to the top-down law enforcement and border-focused efforts of the federal departments and agencies assigned new Homeland Security responsibilities. President Barack Obama has largely continued his predecessor's policies, and congressional oversight has been haphazard.'
23. Stephen Flynn and Jeane Kirkpatrick, 'Written Testimony Before a Hearing of the Committee on Homeland Security and Governmental Affairs, US Senate', *The Department of Homeland Security: The Way Ahead After a Rocky Start* (2005), http://www.cfr.org/publication/7643/department_of_homeland_security.html.
24. Ibid.
25. Ibid.
26. Ibid.
27. Homeland Security Council, *National Strategy for Homeland Security*, 27.
28. Flynn, 'Report Card'.
29. Ibid.
30. Ibid.
31. Ibid.
32. Ibid.
33. Symposium, 'Homeland Security, Law, and Policy through the Lens of Critical Infrastructure and Key Asset Protection', *Jurimetrics Journal* 47 (2007): 259, 260.
34. Department of Homeland Security, *National Infrastructure Protection Plan* (Washington, DC: DHS, 2009), https://www.dhs.gov/sites/default/files/publications/NIPP_Plan.pdf.
35. Lucy L. Thompson, 'Critical Issues in Identity Management – Challenges for Homeland Security', *Jurimetrics Journal* 47 (2007): 335, 348.
36. Ibid., 337.
37. Ibid.
38. Ibid., 333.
39. See Emergency Supplemental Appropriations Act for Defense, the Global War on Terror, and Tsunami Relief, 2005, P.L. 109–13 ss 201–207, 119 Stat. 231, 312–16 (2005).
40. Ibid., 343.
41. Ibid., 347.
42. See Federal Financing Institutions Examination Council, http://www.ffiec.gov/default.htm (accessed December 17, 2013).
43. Ibid., 346. See also FFIEC Central Repository's Public Data Distribution website for compliance reports, https://cdr.ffiec.gov/public/.
44. Flynn, 'Report Card'.
45. Flynn & Kirkpatrick, 'Written Testimony'.
46. Homeland Security Council, *National Strategy for Homeland Security*, 18.
47. Flynn & Kirkpatrick, 'Written Testimony'.
48. Homeland Security Council, *National Strategy for Homeland Security*, 18.
49. Ibid.
50. Light and Lindsay, 'Views of Homeland Security'.
51. Homeland Security Council, *National Strategy for Homeland Security*, 31.
52. Flynn, 'Report Card'.
53. Guiora, *Global Perspectives on Counterterrorism*, 14.
54. Ibid., 15, quoting Alan Cullison, 'Inside Al Qaeda's Harddrive', *Atlantic Monthly*, Sept. 2004.
55. Benjamin Franklin, Pennsylvania Assembly, 'Reply to the Governor, 11 November 1755', in *The Papers of Benjamin Franklin*, ed. Leonard W. Labaree, vol. 6 (New Haven: Yale University Press, 1963), 242.
56. Guiora, *Global Perspectives on Counterterrorism*, 63.

57. The Geneva Convention states that minimising collateral damage is a requirement of international law and nations must limit collateral damage in times of war. Protocol Additional to the Geneva Conventions of 12 August 1949, and relating to the Protection of Victims of International Armed Conflicts (Protocol 1) art. 57 (2).
58. Guiora, *Global Perspectives on Counterterrorism*, 19.
59. Ibid., 242.
60. Sec. 216, H.R. 1, Implementing the 9/11 Commission Recommendations Act of 2007, http://www.cfr.org/content/publications/attachments/911.pdf (accessed December 17, 2013). The Assistant Secretary for Infrastructure Protection must compile an annual report notifying Congress of the following: (1) Changes in the infrastructure vulnerability from the year before; (2) Explanation of the greatest risks facing the country; (3) Recommendations to mitigate those risks.
61. 'One example might be dissemination of preparedness information about potential threats and emergency plans. The state of California currently has a law pending that would require the State Department of Education to electronically distribute disaster preparedness educational materials and lesson plans that are currently available to local education agencies.' Amos Guiora and Kyle McKenzie, 'A Framework for Evaluating Counterterrorism Regulations', *Mercatus Policy Series*, Policy Resource No. 3 (2006): 25–26, quoting Assembly Bill No. 103, California Legislature 2005–06 Regular Sessions, Legislative Counsel's Digest, last amended 22 May 2006, http://www.homeland.ca.gov/legislative.html.
62. 9/11 Commission, *Final Report of the National Commission on Terrorist Attacks upon the United States* (Washington, DC: US Government Printing Office, 2004), 364.
63. Ibid., 364–5.
64. Guiora, *Global Perspectives on Counterterrorism*, 75.
65. See Public Law 107–40, Authorization for Use of Military Force (September 18, 2001), http://www.gpo.gov/fdsys/pkg/PLAW-107publ40/pdf/PLAW-107publ40.pdf.
66. Suzanne Uniacke, 'On Getting One's Retaliation in First', in *Preemption: Military Action and Moral Justification*, eds. Henry Shue and David Rodin (Oxford: Oxford University Press 2007), 88.
67. Homeland Security Act of 2002, 6 USC 101.
68. See Office of Management and Budget, 'Program Assessment Rating Tool (PART)', http://www.whitehouse.gov/omb/performance_past (accessed December 17, 2013).
69. DHS, 'PART Program Ratings', http://georgewbush-whitehouse.archives.gov/omb/expectmore/agency/024.html; and http://www.dhs.gov/xlibrary/assets/cfo_aprappc_fy2008.pdf (accessed December 17, 2013).
70. Ibid.
71. Pam Fessler, 'Experts Challenge Homeland Security Strategy', http://www.npr.org/templates/story/story.php?storyId=18118652 (accessed October 28, 2013).
72. Program Assessment, FBI Counterterrorism Program, http://www.whitehouse.gov/omb/expectmore/summary/10003802.2005.html (accessed October 28, 2013).
73. Ibid.
74. Testimony of Robert S. Mueller, III, Director, FBI, Before the Select Committee on Intelligence of the United States Senate, February 24, 2004, 'Success in The War on Terrorism', http://www.au.af.mil/au/awc/awcgate/fbi/mueller022404.htm.
75. Program Assessment, FBI Counterterrorism Program.
76. Martin S. Feldstein, 'Designing Institutions to Deal with Terrorism in the United States', *American Economic Review* 98, no. 2 (2008): 122–6, http://nrs.harvard.edu/urn-3:HUL.InstRepos:2920114; and *Designing Institutions to Deal with Terrorism in the United States*, National Bureau of Economic Research (NBER) Working Paper 137299 (Washington, DC: NBER, 2008), http://www.nber.org/papers/w13729.
77. Ibid., 10–11.
78. Program Assessment, FBI Counterterrorism Program.
79. Program Assessment, Transportation Security Administration, http://www.whitehouse.gov/omb/expectmore/summary/10003632.2006.html (accessed October 30, 2013).
80. Ibid; and Program Assessment, FBI Counterterrorism Program.
81. Program Assessment, Transportation Security Administration.
82. Statement of Gerald L. Dillingham to the National Commission on Terrorist Attacks Upon the United States, April 1, 2003, http://govinfo.library.unt.edu/911/hearings/hearing1/witness_dillingham.htm.

83. CIA, 'What We Do', https://www.cia.gov/about-cia/todays-cia/what-we-do/index.html (accessed October 30, 2013).
84. Ibid.
85. Ibid.
86. See, for example, *El-Masri v. Former Yugoslav Republic of Macedonia*, App no 39630/09, 13 December 2012.
87. OIG, *Report on Accountability of CIA with Respect to the 9/11 Attacks* (Washington, DC: US Congress, 2005), Executive Summary, http://www.fas.org/irp/cia/product/oig-911.pdf.
88. 10 U.S.C.A. s 111.
89. U.S.C.A Art. 1 s 8 cl. 11.
90. 10 U.S.C.A. s 372.
91. See http://www.whitehouse.gov/omb/expectmore/rating/all.html.
92. See also http://oig.state.gov/documents/organization/105953.pdf.
93. Department of State, *FY 2008 Performance Report*, http://www.state.gov/documents/organization/114172.pdf (accessed October 25, 2013).
94. See, further, the paper by Mueller and Stewart in this issue.

Index

Aboriginals 46
abuse 46, 77, 81
acceptable risk 6, 109–13, 116–17
accountability 2, 4–6, 61, 67, 88; homeland security 122–43; matrix 133–5, 140; UK 92, 95, 100
Accountability Review Team 138
Acts of Parliament 79–80, 82, 84, 88
Addressing Lessons from the Emergency Response to the 7th July 2005 London Bombings 31, 50, 98
Afghanistan 15, 97
air raids 43
airborne capacity 15
Alberta 51
Allen, G. 83
American Civil War 42
anarchists 11, 19
angary 78
Animal Health Act 103
animal rights 12
anomie 2
Anti-terrorism Assistance Program 139
Appleton, C. 3, 9–24
Arab Spring 85
Archaic prerogative powers 78
armed forces 48, 76, 78–9, 82–4, 86–7, 96, 110
Armed Forces Act 84
arrest powers 52
Aryan Nations 12
Asia 68
Attorney Generals 28–30, 78
AUF 9
Australia 3–4, 10, 12, 25–39, 68, 113, 116
Australian Capital Territory 26–7, 33
Australian Defence Force 28
Australian Local Government Association 28
Australian Productivity Commission 34–5
authority figures 64–5
Authorization for Use of Military Force (AUMF) 133
Automated Commercial Environment 128
avalanches 41
Backbench Business Committee 81–2
Bagehot, W. 79–80
Bakker, E. 3, 13, 17
Bakunin, M. 11
balance of rights 4, 16, 53, 122–4, 127–36, 138–9
Baltimore 137
Banks, D. 118
Barriers to Effective Climate Change Adaptation 34
beach erosion 35
Beam, L. 12
Beatty, P. 47
Bell JJ 32
Bellwin Scheme 100
Beredskapstroppen 13
Bills of Rights 44–5, 47, 79
Bin Laden, O. 129
biometrics 127
Birmingham 95
black market 51
black swan occurrences 3, 13
Blair, T. 83, 87, 104
Blick, A. 4, 76–93
Bloomberg, M. 118
Boin, A. 13
bombings 11, 13, 15–16, 18–19, 33; Canada 43; homeland security 122; UK 66, 95–6, 98, 104; US 116
border controls 126
Boston Marathon 122
Brazier, R. 82
break-even analysis 114
Breivik, A.B. 3, 9–24
Breivik Report 3, 10
Brennan CJ 31–2
Bristol 95
Britain 93, 113
British Columbia 43, 50, 52
British Empire 79
British North America Act 41–2
Broderick, J. 59, 61–2
Bronze Command 95
Brown, G. 84, 86–7, 97

bureaucracy 15, 116, 118
Burton, F. 10, 17
Bush, G.W. 6, 122–3, 125–6, 128
Bushfire Management Committee 35
bushfires 33, 35
business continuity plans 51
Butler Inquiry 87

Cabinet Manual 86–7
Cabinet Office Briefing Rooms (COBR) 31, 50, 60, 95, 97–9
Cabinet Office Sector Resilience Plan for Critical National Infrastructure 97
Cabinets 14, 30, 33, 46, 60, 64, 94–5, 98, 104
Cameron, D. 85
Canada 3–4, 12, 28, 40–58, 113, 116
Canada Charter of Rights and Freedoms 46–7, 50
Canadian Border Services Agency 46
Canadian Disaster Database 40
Canadian Firearms Centre 46
Canadian Security and Intelligence Service 46
carcinogens 113
Carlile, Lord 101
catastrophic events 25, 29–30, 36, 99, 115, 125
Category 1 responders 60–1, 63–4, 67, 96
Category 2 responders 60, 66, 96
censorship 42
Central Intelligence Agency (CIA) 138
centralisation 59, 67
Chertoff, M. 99, 137
Chilcot Inquiry 87
China 44
Christensen, T. 13–14, 16
Cincinnati 137
citizenship 126
civil contingencies *see* contingency planning
Civil Contingencies Act (CCA) 3–6, 28, 31, 34, 59–75, 82–3, 88, 92–104
Civil Contingencies Enhancement Programme (CCAEP) 98
Civil Contingencies Secretariat (CCS) 98
Civil Contingency Reaction Forces (CCRF) 96
Civil Defence Act 48
Civil Emergency Measures Act 52
Civil Protection Mechanism 2
civil rights/liberties 4, 16, 18–19, 40–1, 45–8; balancing 124–5, 128–35; Canada 50–3; homeland security 124–5, 128–31, 133–4, 138
Civil Service 84
Civil Service Act 83–4, 88
Civil Service Code 81
civil society 95
climate change 36, 82
club government 98
Co-operative Committee on Japanese Canadians 44
Coast Guard 128
Cold War 4, 40–1, 44, 48, 82
collateral damage 128, 131, 134–5
colonialism 26
command structures 4, 59, 61–3, 68–9, 95, 99–100, 104
command-and-control 59, 61, 66, 68–9, 95, 99–100
commandeering of property 34, 50
9/11 Commission 127, 132–3, 137
common law 31, 77
Commonwealth 4, 25–33, 36
Commonwealth Disaster Plan (COMDISPLAN) 29–30
communication 1, 12, 15, 18, 30; Canada 42, 51, 53; homeland security 127–8, 138; risk 6, 109–10, 115–16; Royal Prerogative 81, 88; UK 62–6, 68–9, 95, 102; US 118
Communists 43
community 2, 35–6, 50, 53, 63, 65, 125
Community Risk Register 60
compensation 51
conduit media 66–7
Congress 83, 123, 131–40
conscription 48, 51
Conservative Party 86
Consolidated Revenue 32
Constitutional Reform and Governance Act 84, 87
Constitutional Renewal Bill 84
constitutionalism 1–8
Constitutions 26–8, 32–3, 41–2, 76–80, 82–6; Royal Prerogative 88; UK 97; US 109, 140
Container Security Initiative 126, 128
contexts 1–3
contingency planning 1–10, 13, 15–16, 19, 25–39, 63, 92–104, 118
Continuation of Transitional Powers Act 44
Cook, R. 86
Copeland, D. 11
copycat behaviour 17
copyright 69
Correctional Service of Canada 46
cost-benefit analysis 6, 109–10, 113–18, 133–6, 139
costs per saved life 6, 109–11, 116
Council of Australian Governments (COAG) 27–8, 36
Council of Europe *see* European Council
counter-terrorism 3, 5–6, 9–10, 13–14, 16–19; Australia 29; homeland security 122, 124–5, 133; UK 98–9, 102–3; US 109–21
Court of Session 94
Crennan JJ 32
crisis management 3, 9–10, 13–16, 19
Critical National Infrastructure 97
critical service providers 51
Cross, J. 45
Crown 31, 77–8, 84

Crown Prosecution Service 100
Cuban Missile Crisis 44, 48
culture of fear 93
curfews 50, 102
Curtis, A. 12
customs 16, 126, 128
Customs-Trade Partnership Against Terrorism (C-TPAT) 128
cyber-security 45, 59, 87, 127
Cyclone Tracey 32

Dare, General 45
Darwin 32
Davis v. Commonwealth 31
De Graaf, B. 3, 13, 17
decentralisation 59, 62
decision-making 6, 13, 46, 53, 109–10; homeland security 131, 135; US 113, 115–16, 118
Defence of Canada Regulations 45, 50, 52
Defence Committee 96
Delta force team 13–15
democracy 2–3, 5, 9, 13–14, 16; Canada 44; homeland security 129, 133, 140; Norway 18–20; Royal Prerogative 77–8, 80, 84, 87–8; UK 69
demolition of property 50
Department of Defense, US 123–4, 139
Department for Environment, Food and Rural Affairs (DEFRA), UK 98–9
Department of Homeland Security (DHS) 3, 6, 28, 46, 99, 110, 116–17, 122–3, 136–7
Department of Public Safety and Emergency Preparedness Act 46
Department of State, US 139
Department of Transportation, US 138
deportation 42, 44, 50, 102
Detention of Terrorist Suspects Bill 5
devolution 3, 82, 93
Dicey, A.V. 77
digitisation 62, 69
Dillingham, G.L. 138
Director General of the Security Service 103
Director of Public Prosecutions 103
Directorate of the Police, Norway 14
disaster resistant communities 53
disease 1, 41, 50, 92–3, 99
displacement 1
disposition of property 50
Doomsday powers 101
Draft Detention of Terrorist Suspects Bill 92, 100, 103–4
drones 3, 87

Eburn, M. 4, 25–39
effectiveness 2, 5–6, 17, 42, 84; homeland security 122–3, 128–9, 132–3, 135–40; US 113–16, 118
emergencies 1–6, 15–17, 19, 28–36, 59; Australia 25; Canada 40–58; definitions 45–6, 93, 101, 103; Royal Prerogative 76, 87–8; UK 60, 92–108; US 136
Emergencies Act 40, 45–8, 50–3
Emergency Management Act 46
Emergency Management Australia (EMA) 29–30, 32
Emergency Management and Civil Protection Act 49
Emergency Management Liaison Officers 29
Emergency Measures Act 48, 51
emergency powers 3–5, 31, 33, 40–1, 44; Canada 48–9, 51, 53; Royal Prerogative 76–91; UK 59, 61, 92–3, 96, 98–9
Emergency Powers Act 44, 93, 96
Emergency Preparedness 65, 67–8, 96
Emergency Preparedness Act 45
Emergency Response and Recovery 94–6, 98
Emergency Support Teams 28
Emergency Workers Act 97
Energy Act 103
England 12, 65, 77, 79, 88, 94
English Civil War 61
The English Constitution 79
English language 63
Enhanced Terrorism Prevention and Investigations Measures Bill (ETPIMs) 5, 92, 102–4
Environment Agency 99
Environmental Protection Agency 117
espionage 43
ESRC 6
ethics 3–4, 6
Europe 10, 12, 79, 88
European Act 80
European Convention on Human Rights 87, 102
European Council 2, 80
European Union (EU) 2, 80
Europol 10, 17
Euthanasia Laws Act 27
evacuation 34, 44, 50, 63–5
Executive Orders 124
executive power 2–3, 5, 9, 14, 30–4; homeland security 123–6, 131, 133, 136, 139–40; Royal Prerogative 76–9, 81, 83–4, 86–8; UK 97, 100, 102, 104
Exercise Watermark 69
extremism 11, 14, 17–18

Facebook 68
Fallows, J. 116
Fascists 43
fatality rates 111–13
fear-mongering 118
Federal Bureau of Investigation (FBI) 137, 139
federal coordinating officers 28
Federal Emergencies Act 46, 50

Federal Emergency Management Agency (FEMA) 46, 128
Federal Financial Institutions Examination Council (FFIEC) 127
Feldman, M. 11
Fimreite, A.L. 14
financial crisis 32–3
fining powers 52
fire brigades 34–5, 60, 97, 110, 125
Fire Control Officer 34–5
fires 25, 30, 33, 35–6, 51
First Fleet 26
First World War 4, 40–3, 46, 53, 79–80
fiscal responsibilities 32, 124, 128, 131–2
Fixed-term Parliaments Act 86–7
floods 1, 25, 33, 36, 48; UK 59, 65–6, 68, 98–9, 104; US 117
Food and Drugs Act 46
Food and Environment Protection Act 103
forecasting 3, 13
Foreign Affairs Committee 80–1
Foreign Intelligence Surveillance Act (FISA) 125
Fort Hood 12
Fortuyn, P. 12
France 12
Franklin, B. 129
Freedom of Information Act 82
freedom of speech 17, 19
French CJ 32
Front de libération du Québec (FLQ) 41, 44–6, 53
fuel protests 82, 93
Furedi, F. 19, 93
future generations 115

Gable, G. 11
Geis, D. 41, 53
General Accounting Office (GAO) 138
geopolitics 85, 129, 131–2, 134–5, 138–9
Germany 12
globalisation 1
Gold Command 60, 64–5, 68, 95
Goldwater Nichols Act 139
good practice 60, 95
governance 1–2, 5, 26–8, 36, 62, 84, 92–108
The Governance of Britain 84–5, 87, 97
Government Communications Headquarters (GCHQ) 81
Governor in Council 42, 45, 47
Guantanamo Bay 3
Guiora, A. 6, 122–43
Gulf War 82
Gummow JJ 32

habeas corpus 47
hackers 45
Hague, W. 86–7
Hasan, N. 12
Hayden, W. 33
hegemony 61
Her Majesty in Council 93
heterarchy 61–2, 99–100
Hewitt, C. 10, 12
hierarchy 59, 61–4, 100
High Courts 26–7, 94
Hilton Hotel 33
history 41–6, 48, 53, 113
Hobbes, T. 2, 61, 109
Home Affairs Committee 97
Home Office 31, 50, 98, 100, 102, 104
Home Secretaries 95, 102
homeland security 3, 6, 28, 46, 99, 110, 114–17, 122–40
Homeland Security Act 136
House of Lords 83–5, 87, 104
Houses of Commons 79, 83–7, 96, 166
human rights 2–3, 9, 18–19, 53
Human Rights Act 82, 87, 93–4
Humanitarian Assistance Centre 95
Hurricane Katrina 3, 67, 93
hurricanes 1, 3, 67, 93
Hutton Inquiry 87

identity management systems 127
ideology 10–11, 14, 17–18, 43, 124
immigrants 12, 42, 48, 126
Immigration and Naturalization Service 126
Independent Reviewer of Terrorism Legislation 103
individual liberties *see* civil rights/liberties
informants 17
Information Age 59, 63
information gathering 18
Information Society 62–3, 189
information-related provision 59–75
infrastructure 2, 17, 19, 30, 45; homeland security 123–8, 135–6, 138; UK 97, 99–100
Inquiries Act 84
Institute for Safety, Security and Crisis Management (COT) 12
Intelligence and Security Committee 81, 103
intelligence services 16–19, 81, 125, 131, 134–5, 137–9
Intelligence Services Act 81
Intelligence Services Commissioner 103
Intelligence-led Policing (ILP) 125
interactivity 69
Interception of Communications Act 81
interdisciplinarity 59
International Covenant on Civil and Political Rights 47, 51–3
international law 53, 135
internet 10, 12, 18, 67, 127–8
internment 43–4, 47
Iraq War 5, 76, 83, 85–7, 97, 104

Ireland 77, 93–4, 113
Irvine, D. 10
Islamism 11, 110, 137
Israel 137
Italy 12

Jackson, P. 11
James I/VI, King 78–9, 81
Japanese Canadians 43–4, 47
Jews 12
Joint Chiefs of Staff 139
Joint Committee on the Draft Detention of Terrorist Suspects Bill 100, 103
Joint Inquiry (JI) Report 138
Joint Rescue Coordination Centres, Norway 14
Judicial Committee of the Privy Council 43–4
judicial reviews 81, 135
judiciary 78–9, 133, 136
July 7 2005 66, 92, 95, 98, 104
22 July Commission 10, 14–17
Justice and Security Act 81

Kaplan, J.J. 12
Korean War 4, 40, 44, 48
Kosovo 82
Kristiansen, J. 16
Ku Klux Klan 12
Kunreuther, H. 116

Labour Party 9, 19, 82–4, 86, 103
Labrador 50
Lackawanna Group 122
Lagdec, P. 13
Laporte, P. 45
Lapshyn, P. 12
Lash, S. 62–3
Law Lords 81
Lead Government Department (LGD) 60, 95, 98–9
leaderless resistance 12
leadership 3, 17, 28, 30, 59–61, 63–5, 67, 69, 137, 139
Lebanon 116
Lee, H.P. 30–1, 33
Leeds 95
Legal Prerogatives of the Crown 78
legislation 2–5, 16, 27–8, 30–1, 33–4; Canada 40–58; homeland security 122–3; Royal Prerogative 78–83, 86–8; UK 59, 93, 95, 97–9, 101, 103–4; US 139
legislature 42, 78–80, 82, 87, 132–3
Lester, Lord 82
Leviathan 61, 109
Liaison Committee 81
liberal democracy 2, 129, 133
Libya 86–7
Lindsay, J. 4–5, 40–58
Little Rock, Arkansas 11
livestock 50
lobbyists 110
Local Emergency Management Committee (LEMC) 34
local government 34–5
Local Resilience Forums (LRFs) 60, 66, 68, 99
Lockerbie 116
Lomborg, B. 117
London 11, 66, 95
London Underground 66
lone wolf terrorists 3, 9–24, 122–3

McBain, G. 87
McCormack, W. 124
Macdonald, Lord 100–2
McGuinness, M. 6
McVeigh, T. 11
magistrates 94
Mali 86
manifestos 12
Manitoba 48, 50–2
Marriage Act 27
martial law 87
matrix for accountability 133–5, 140
media 4, 12, 59, 63–9, 124, 133
Megaports Initiative 128
mental health 17
Merah, M. 12
Metzger, T. 12
MI5 81
MI6 81
Miami 122
Miliband, D. 2
militancy 17
military 44–5, 77, 79, 82–3, 85–7; UK 94, 96–7, 100; US 123–5, 131, 133, 139
Military Aid in the Civil Community (MACC) 96
Military Assistance to the Civil Authorities (MACA) 96
millennium 41, 45–6
Ministerial Code 81
Ministerial prerogative powers 78
Ministers of Public Safety and Emergency Preparedness, Canada 28
Ministry of Defence, UK 96
Ministry of Justice, Norway 14, 16
Ministry of Justice, UK 78, 88, 97
Minneapolis 137
Montauban 12
Moosavian, R. 3–4, 59–75, 93–5
Mueller, J. 5, 109–21
Mueller, R.S. 137
Muhammad, A.M. 11
multiculturalism 12
Municipal Coordinator 35
Municipal Emergency Management Committee 35

Municipal Emergency Management Plan 35
Municipal Emergency Planning Committee 35
murder 9, 12, 44
Muslims 12

National Academy of Sciences, US 116
National Association of Japanese Canadians 47
National Catastrophic Natural Disaster Plan (NATCATDISPLAN), Australia 29
National Commission on Terrorist Attacks Upon the United States 138
National Counter-Terrorism Plan, Australia 29
National Emergency Transitional Powers Act 44
National Infrastructure Protection Plan, US 127
National Response Plan, US 128
National Risk Register, UK 60, 64
National Security Strategy, UK 62
National Strategy for Disaster Resilience, Australia 25, 36
natural disasters 28–36, 40–1, 48, 52, 59; homeland security 125, 128, 136; UK 68, 93; US 117
Natural Disasters Organisation, Australia 32
Netherlands 12
New Brunswick 51
New South Wales 34
New York 118
New Zealand Law Commission 30–1, 33
Newark 137
Newfoundland 50
News Co-ordination Centre 64
Nichols, T. 11
Nordre Buskerud 14
Northern Ireland 93–4, 113
Northern Territory 26–7, 32, 35
Northwest Territory 51–2
Norway 3, 9–24
Nova Scotia 51–2
Nuclear Regulatory Commission 117
Nunavut 50–1

Obama, B. 2, 6, 10, 122–3
Office of the Coordinator for Counterterrorism 139
Office of Critical Infrastructure and Emergency Preparedness (OCIPEP) 45–6
Office of Homeland Security 124, 127
Office of Inspector General (OIG) 138
Office of Management and Budget (OMB) 115, 136–9
Oklahoma City 11
Olympic Park 11
Omar, M. 129
Ontario 49–53
Operation Iraqi Freedom 137
Operations in the UK 96
Orders in Council 42–4
organisational models 59
Oslo 9–10, 12–15
Oslo District Court 9
overreaction 115–16, 118
overview 1–8

Pakistan 137
Palestine Islamic Jihad 137
pandemics 1
panic 64
Pantucci, R. 11
Pape's case 33
Parliaments 5, 10, 13, 26, 42; Canada 44–5, 47; Royal Prerogative 76, 78–88; UK 94–5, 97, 100–1, 103–4
parochialism 99
Passachendaele, battle of 42
passports 78, 83, 87, 116, 127
Paté-Cornell, E. 110, 118
patriarchy 61
patriotism 12
Pearl Habor 43
Phillips, L. 44
Pine Lake Alberta 41
Pistole, J. 117
Pitt Review 60, 62, 65, 98–9
POGG powers 42–3
Poland 16
Police Security Service, Norway 14, 16
policy-making 1, 5–6, 80, 109–21, 138
pollution 113
Ponsonby, A. 79–80
Ponsonby Rule 80–2, 84–5
population growth 36
ports 128
Posner, R. 137
power to react 40–59
Prevention of Terrorism Act 102
price fixing 51
Prince Edward Island 50–1
prisoners of war 43
private sector 1, 29, 36, 59–60, 100, 123, 125–7
privatisation 93
Privy Council 43–4
Program Assessment Rating Tool (PART) 136–7
property controls 50
Protect America Act 125
Protection of Freedoms Act 100–1
Protestants 79
Provincial Acts 46, 48
provincial legislation 44
psychology 64, 110
Public Administration Select Committee (PASC) 83
public gatherings 50
public information 59–75
Public Safety Act (PSA) 45–6
Public Safety Canada 40, 46
public sector 1, 59, 100

Al-Qaeda 122, 125, 133, 137–8
qualified service 51–2
Quebec 44–5, 50
Queensland 35
Questions of Procedure for Ministers 81

racism 12
Radcliffe, Viscount 31
radio 66, 68
Railways Act 103
REAL ID Act 127
recommendations 16, 83, 98, 137
Recovery Co-ordinating Working Group 95
Regional Civil Contingencies Committee (RCCC) 94, 98
Regional Controllers 34
Regional Emergency Committee 35
Regional Emergency Coordinators 34
Regional Emergency Management Committee 34
Regional Emergency Operations Controller 34
Regional Nominated Coordinator (RNC) 94
Regional Recovery Committee 35
Regional Recovery Teams 94
Registration of Alien Enemies 43
regulation of essential resources 50–1
Regulations for the Defence of Canada 43–4
Reid, R. 122
Renfree, H.E. 31
Report on Accountability 138
reporting frequency 131–2
requisition of property 31, 50, 98
Reserve Forces Act 96
Reserve Reaction Force 96
resilience 1–8, 19, 25, 36, 60; homeland security 129, 133; UK 31, 50, 62, 92, 95, 97–9, 104
Responders 4, 60–1, 63–4, 66–7, 96
responsibilities 28–35, 109–21, 128, 131–2
Responsibility to Protect 82
7/7 Review Committee Report 66
Review of Counter Terrorism and Security Powers 100, 102
Rights of the Terminally Ill Act 27
risk analysis/management 6, 28, 60, 92–3, 95; homeland security 124, 127; UK 98–100, 102; US 109, 115–18
risk society 1, 62
Robinson, L. 110
Rose Marches 14
Royal Assent 78, 101
Royal Canadian Mounted Police 46
Royal National Lifeboat Institution 97
Royal Navy 78, 84, 87
Royal Prerogative 4–5, 31, 76–93, 97
Royal United Services Institute 6
Rudolph, E. 11
rule of law 2–3, 122, 131–2, 135–6
Rural Fire Service 34
Russia 129

sabotage 50
Saskatchewan 51
scanners 115, 117
Scotland 3, 77, 79, 93–4, 116
search powers 52
Second World War 4, 11, 13, 40–1, 43–4, 46–8, 50, 53, 80
Secret Intelligence Service 81
Secure Freight Initiative 128
security 1–2, 6, 10, 12–19, 41; Canada 45, 48; definitions 122–40; Royal Prerogative 76–7, 80–2, 84, 88; UK 61, 93, 95, 99, 101–2; US 110, 113–18, 122
Security Council 85–6
Security Intelligence Organisation 10
Security Service 103
Security Service Act 81
Select Committee on the Constitution 83, 87, 104
Select Committee on Intelligence 137–8
select committees 81–3, 85, 87, 104, 137–8
Senate 137–8
September 11 2001 2, 6, 110, 114, 116; Canada 41, 45; homeland security 122–4, 127–8, 138–9; Norway 14; UK 82–3, 93
severe acute respiratory syndrome (SARS) 41
sheriffs 94
Shone, A. 18
short-termism 129
Siam 78
Silver Command 95
Simon, J.D. 10
Smith, J. 103
social media/networks 18, 68–9
social networking sites (SNS) 68–9
Social Security 126
Solicitor General 46
Sorensen, J. 66
South Asia Military Assistance Program 139
South Australia 35
sovereignty 2, 46, 124
Spaaij, R. 10–13, 17–18
Spain 12
Spanish Civil War 43
special measures 3, 96
St-Laurent, L. 44
state 1–3, 16, 26–7, 41, 44; UK 62, 93, 95, 97, 99–100; US 123–4
state coordinating officers 29
State Emergency Services (SES) 33–5
states, Australia 28–30, 32–6
states of emergency 34, 40–1, 46–9, 52–3, 93, 101
Stewart, M. 5, 109–21
Stewart, S. 10

Stoltenberg, J. 14, 19
Strategic Coordinating Groups (SCG) 60, 95
Strategic Policy Institute 30
Strategy for Homeland Defense and Civil Support 124
Straw, J. 86
Stretton, A. 33
strikes 93–4
Student and Exchange Visitor Information System (SEVIS) 126
Suez Canal 80
Supreme Courts 42
surveillance 17–18, 102, 125
Syria 85–6

Tasmania 26, 35
teamwork 16
technology 62–3, 66, 68–70, 93, 117, 125–6, 138
television 66, 68
Territorial Army 96
terrorism 1–3, 5–6, 9–24, 28–30, 41; Canada 44; homeland security 122–40; UK 59, 82, 92–3, 95, 97–103; US 109–21
Terrorism Act 100–1
Terrorism Prevention and Investigation Measures (TPIM) Act 102–4
Terrorist Interdiction Program 139
theory 61–2
threats to homeland security 1, 5, 9–10, 12–14, 17; Australia 36; Canada 45–6, 48, 52; Royal Prerogative 77, 82; UK 62, 93, 97, 101, 103; US 113–16, 118, 122–4, 127, 129–32, 135–9
Times Square 115, 122
tornadoes 41, 117
Toronto 122
totalitarianism 2
Toulouse 12
trade 42–3, 78, 128
trade unions 81
traffic accidents 113–14
transparency 2, 16, 20
Transportation Security Administration (TSA) 137–8
travel regulations 49–50
trespassing 50
Triple Lock strategy 94, 101
Trudeau, P. 45
tsunamis 33, 68
Twigg, J. 61, 67, 69
Twitter 68
Typhoon Haiy 93

unintended consequences 1, 17
Union of Democratic Control (UDC) 79–81, 85
United Kingdom Operations (UK Ops) 97
United Kingdom (UK) 2–3, 5, 26, 42–3, 62; acceptable risk 113, 116; and Australia 28, 31–2, 34; governance 92–3, 100–1; homeland security 122; Royal Prerogative 76–80, 82–3, 85–6, 88
United Nations (UN) 51, 85–6
United States (US) 2–3, 5–6, 10–12, 14, 28; and Canada 42, 46; policy-making 109–21; and UK 83, 85, 100
Unmanned Aerial Vehicles (UAVs) 87
user-generated content (UGC) 68
Utøya 9–10, 12–15, 19
Utvika 14

values 2, 13–14, 19–20, 63, 125, 131
Van der Graaf, V. 12
Victoria 35
Vimy Ridge, battle of 42

Wales 3, 77, 93–4
Walker, C. 1–8, 59, 61–2, 92–108
War Measures Act 4, 40, 42–4, 46–7, 53
war powers 83, 85–7, 97, 139
War Powers Act 83
war on terror 2, 44, 137, 139
warning systems 60, 63–6, 114
Washington 122
watch lists 16
weapons controls 16, 18
weapons of mass destruction (WMD) 125
websites 16
West 9, 11–13, 19, 113
Western Australia 35
White House 133
white supremacists 11–12
Whitehall 84
Whitty, G. 61
Wildavsky, A. 6
Wildfires Act 51
Worldwide Security Upgrades Program 139
Wright, T. 83

Y2K 41, 45–6
Young, G. 86
Yukon 51–2

Zedner, L. 18

For Product Safety Concerns and Information please contact our EU representative GPSR@taylorandfrancis.com Taylor & Francis Verlag GmbH, Kaufingerstraße 24, 80331 München, Germany

Batch number: 10726294

Printed by Printforce, the Netherlands